Conveyors

Pat has created a complete guide for material handling conveyors and basic automation. This is a "must have" resource for anyone in our business. The book provides the concepts and tools for understanding and improving the material flow that drives our economy behind the scenes. We've issued his book to all of our sales engineering and business development staff.

-Gary Cline, CEO / Material Handling Executive

This new edition continues on its journey of being an introduction to the wide variety of conveyor types and application methods available. It discusses conveyor and system design and in its new edition, expands on information about sorters and motorized roller conveyors, new information on robotic interfaces, as well as some of the trade-offs made when conveyor design rules are intentionally broken. Each chapter now also includes questions and problems to help reinforce the material covered.

Conveyors: Application, Selection, and Integration includes a pictorial of various installations of conveyors in detail that the reader can model after. It presents common application mistakes made and discusses how the reader can avoid them in the future during the integration process. The book focuses on conveyor application information that is offered in just this book and discusses the effect that extreme temperatures have while providing real-life examples of conveyor system design that may at times break common design rules.

This book is a resource for engineers, managers, and executives who want to gain a better understanding of conveyors and their application. It is also for colleges and universities that offer industrial and/or manufacturing engineering curricula.

Systems Innovation Book Series
Series Editor: Adedeji Badiru

Systems Innovation refers to all aspects of developing and deploying new technology, methodology, techniques, and best practices in advancing industrial production and economic development. This entails such topics as product design and development, entrepreneurship, global trade, environmental consciousness, operations and logistics, introduction and management of technology, collaborative system design, and product commercialization. Industrial innovation suggests breaking away from the traditional approaches to industrial production. It encourages the marriage of systems science, management principles, and technology implementation. Particular focus will be the impact of modern technology on industrial development and industrialization approaches, particularly for developing economics. The series will also cover how emerging technologies and entrepreneurship are essential for economic development and society advancement.

Innovation Fundamentals
Quantitative and Qualitative Techniques
Adedeji B. Badiru and Gary Lamont

Global Supply Chain
Using Systems Engineering Strategies to Respond to Disruptions
Adedeji B. Badiru

Systems Engineering Using the DEJI Systems Model®
Evaluation, Justification, Integration with Case Studies and Applications
Adedeji B. Badiru

Handbook of Scholarly Publications from the Air Force Institute of Technology (AFIT), Volume 1, 2000-2020
Edited by Adedeji B. Badiru, Frank Ciarallo, and Eric Mbonimpa

Project Management for Scholarly Researchers
Systems, Innovation, and Technologies
Adedeji B. Badiru

Industrial Engineering in Systems Design
Guidelines, Practical Examples, Tools, and Techniques
Brian Peacock and Adedeji B. Badiru

Leadership Matters
An Industrial Engineering Framework for Developing and Sustaining Industry
Adedeji B. Badiru and Melinda Tourangeau

Systems Engineering
Influencing Our Planet and Reengineering Our Actions
Adedeji B. Badiru

Conveyors

Application, Selection, and Integration

Second Edition

Patrick M. McGuire

CRC Press is an imprint of the
Taylor & Francis Group, an **informa** business

Second edition published 2024
by CRC Press
2385 Executive Center Drive, Suite 320, Boca Raton, FL 33431

and by CRC Press
4 Park Square, Milton Park, Abingdon, Oxon, OX14 4RN

CRC Press is an imprint of Taylor & Francis Group, LLC

First edition published by CRC Press 2009

ISBN: 978-1-032-45118-3 (hbk)
ISBN: 978-1-032-45362-0 (pbk)
ISBN: 978-1-003-37661-3 (ebk)

DOI: 10.1201/9781003376613

Typeset in Times LT Std
by SPi Technologies India Pvt Ltd (Straive)

Access the Support Material: www.routledge.com/9781032451183

Contents

Preface

The purpose of this book is to prevent someone from making some of the same mistakes I have made. When I first graduated from college with an associate's degree in manufacturing technology, I had to admit at my very first job interview that I had absolutely no idea what a tabletop chain conveyor was. I had never seen one, read about one, or even heard it mentioned in any of my classes. Needless to say, I have learned a lot since that time, and this book is an effort to share some of that knowledge. This second edition is a testament to the fact you never stop learning because I have more to share.

Put simply, this book is about conveyors. I know that sounds incredibly dull and boring, and it is, unless you're trying to move material from point A to point B without manual intervention. Unless you have conveyor experience, you will need help deciding on the best mode of transportation. This book is meant to be an introduction to the wide variety of conveyor types and application methods. It is not meant to be a comprehensive guide; if it was, it would be several volumes. This book discusses conveyor and system design, but it is not meant to be an engineering manual. What I am trying to accomplish is to introduce the various conveyor types with enough information about them that you can talk more intelligently about conveyors and know some of the questions to ask and what to look out for. This book does not take a myopic view of a specific type of conveyor but rather offers a broad overview of all of the major conveyor types. Instead of covering just bulk material handling or just screw conveyors, this book covers the major types of conveyors and a couple of the minor ones as well. There are always new and unique designs and specialty conveyors that should not be included in such an introduction. An answer key to the questions at the each chapter is available on the Routledge website (www.routledge.com/9781032481183).

It is my sincere desire that everyone learns a little more each time they pick up this book.

Acknowledgments

First and foremost, I need to thank God for all he has blessed me with. I would be desperately remiss if I did not start with the person who has encouraged me through this: Sylvia, my wife. There are many other people over the years who have had an impact on my ability to author a book of this nature. The first such person was Morrell Jorgenson. He was the owner of the first conveyor company I ever worked for and through his desire to fill many customers' needs for unique material handling solutions, I gained a very broad base of experiences. During my tenure at Rapistan, my manager Clair Fairbrother and his boss Paul Todd kept me busy with a wide variety of very unique material handling projects.

About the Author

Patrick M. McGuire has been in the material handling industry for more than 20 years. He has a BS degree in computer-integrated manufacturing systems and a professional engineering license for industrial engineering. He started with a very small conveyor manufacturer in upstate New York, J & S Conveyors, where he was the engineering manager. They designed and built a wide variety of conveyors and manufactured tabletop chain conveyors, flat belt conveyors, heavy unit load conveyors, and troughed belt conveyors. After J & S, he went to work for Rapistan, the world's largest manufacturer of material handling systems, specializing in unit handling, primarily distribution systems. Rapistan has since been bought by Siemens AG. McGuire was involved with the office that also specialized in building in-process manufacturing systems, such as one-of-a-kind conveyors and systems for giants Corning, Kodak, Xerox, General Motors, Ford, and Chrysler. He worked as a senior systems engineer and a project manager.

After 10 years of custom product and system design, he took a position with the products group, where he worked on such things as the high-speed divert for airport baggage after which he became the product manager for the heavy unit load product line. In that position, McGuire led a group of engineers in designing, documenting, and selling a full line of pallet-handling conveyors. He has since worked as the director of product engineering for American Ironhorse Motorcycles, and the director of manufacturing and technology for Transnorm System, a specialty conveyor manufacturer. Following Transnorm, he was the manager of engineering services and director of manufacturing for Glidepath. As a member of CEMA, he co-authored or edited several chapters of the new *CEMA Application Guide for Unit Handling Conveyors*.

After leaving Glidepath, he was director of engineering at Orteq, a company specializing in heavy equipment for the oil and gas industry. From there, he was director of engineering at the Vince Hagan company designing portable and stationary concrete batch plants. There, he designed the only concrete plant to fit in a military transport aircraft for the US military.

Currently, he is a senior project engineer at N. J. Malin and Associates, where he leads and mentors a group of engineers and draftsmen in the art of material handling problem-solving. He and his team have worked with such companies as Mars Wrigley, Ariat, CEVA, Texas Instruments, DHL, UPS, and AT&T delivering unique material handling solutions across the United States and Mexico.

1 Introduction

This book is about conveyors, many types of conveyors. The ensuing pages cover how to select the appropriate type of conveyor based on the materials to be handled and then introduce the reader to each type of conveyor. First, however, we must start with some background information.

1.1 WHAT IS MATERIAL HANDLING?

Conveyors are just one subset of the much larger group of material handling equipment. Through the proper application and use of material handling equipment, we try to minimize or even eliminate the manual handling of material.

Material handling is all about movement; raw materials, parts, boxes, crates, pallets, and luggage must be moved from one place to another, from point A to point B, ideally in the most efficient manner. The material being handled is virtually limitless in size, shape, weight, or form. Material can be moved directly by people lifting and carrying the items or using handcarts, slings, and other handling accessories. Material can also be moved by people using machines such as cranes, forklift trucks, and other lifting devices. Finally, material can be moved using automated equipment specifically designed for mechanically handling the items such as robots and, as discussed in this book, conveyors.

The Material Handling Institute of America (MHIA) offers the following as one definition of material handling:

> Material handling is the art and science associated with providing the right materials to the right place in the right quantities, in the right condition, in the right sequence, in the right orientation, at the right time, at the right cost using the right methods.

1.2 WHAT ARE THE MAJOR OBJECTIVES OF CONVEYOR APPLICATION?

Conveyors, as with all material handling equipment, do not add value to the parts, products, or pieces that are being moved. They do not shape, form, process, or change a product in any way. They are totally processes of service, and as a service they have an indirect bearing on product cost as part of the overhead.

The following is a list of some of the major objectives of implementing conveyors:

- Reduce actual manual handling to a minimum.
- Perform all handling operations at the lowest reasonable cost.
- Eliminate as many manual operations as possible.

DOI: 10.1201/9781003376613-1

- Ease the workload of all operators.
- Improve ergonomic considerations for each operator.
- Improve workflow between operations.
- Provide routing options for intelligent workflow.
- Increase throughput.
- Carry product where it would be unsafe to do so manually.

1.3 WHAT ARE THE REAL COSTS OF MATERIAL HANDLING?

Many people do not initially comprehend the full extent of all the costs included in the multitude of operations involving the handling of material. For example, to move pallets from point A to point B, a forklift might be the solution of choice. Along with the cost of a forklift, the first obvious addition is the cost of the operator's wages and benefits. Less obvious and frequently overlooked are the costs of fuel (propane or electricity), driver training and possible certification, periodic maintenance, catastrophic failure repair of the forklift itself, and insurance against possible damage. These together constitute the real cost of that material handling solution.

The same holds true for conveyor systems. The following is a list of possible costs in addition to the cost of the actual equipment and the associated electrical controls:

- Mechanical and electrical installation.
- Information technology (IT) programming for any interface with a warehouse management system (WMS) or host enterprise resource planning (ERP) system.
- Operator training.
- Periodic maintenance.
- Spare parts.
- Cost of stocking spare parts on site.
- Specialized training for maintenance personnel.
- Maintenance contracts.

Now, of course, if the conveyor consists simply of a couple of 3-m (10-ft) pieces of skate wheel conveyor, virtually this entire list can be disregarded. As the system grows, however, more and more of this list will come into play.

1.4 THE DRIVING FORCE

All of the powered conveyors we discuss use electric motors to drive them, pull the belt, or turn the rollers. Every type of conveyor and every manufacturer have specific drive configuration choices available depending on the application. They all consist of an electric motor and a mechanical gear reducer. Some specialty conveyors are available with pneumatic or hydraulic motors in lieu of electric motors, but we will not be covering them.

Reducers are often referred to as speed reducers or gear reducers, but in this book they are simply called reducers because their primary purpose is to gear down or reduce speed. Reducers come in a wide variety of shapes and sizes, but they all have a few things in common. They all have an input shaft, although some will have a hollow bore for mating to a motor. They all have an output shaft, and again, it might have a hollow bore for mating to the shaft it is driving. Finally, they all have two or more gears inside that mesh together to reduce the speed from the input shaft to the output shaft.

There are three primary types of drive arrangements:

1. C-Face-mounted motor and reducer: A motor mounted to a standard input flange of a mechanical gear reducer.
2. Separate motor and reducer: A motor mounted separately from the reducer but connected to the reducer through v-belts and sheaves or chains and sprockets.
3. Gearmotor: An integral motor and reducer.

We are not going into all of the details of every variety of drive configuration available. Figure 1.1 shows two types of motors. The one on the left is referred to as a C-Face mounted motor because it is mounted using the four bolts in the face of the motor. There are several standard-size designations for C-Face mountings. The motor on the right is a foot-mounted motor because it is mounted using the foot plate.

There are also other characteristics to pay attention to when selecting a motor. The environment in which it will be used will play a significant role. There are open motors, which allow free airflow through the motor for cooling. But if it is a dusty environment, you don't want the dust getting into the motor. So, you have the choices of totally enclosed fan cooled (TEFC) or totally enclosed non-ventilated (TENV) motors. There are also explosion-proof motors for use in areas where the environment contains combustible materials. Paper plants are an excellent example because paper dust is highly flammable so any spark could ignite it. Some motors are designed to be washed down in food-handling areas. Other motors are designed for use outside in the elements. All these things need to be considered.

FIGURE 1.1 Motors.

TABLE 1.1
Drive Train Comparison

	Roller Chain	V-Belt & Sheaves	Timing Belt
Noise	Moderate	Low	Moderate
Requires Periodic Lubrication	Yes	No	No
Shock Load Isolation	Low	Moderate	High
Efficiency	95%	93% (88% if not properly tensioned)	98%

Because most motors spin very fast at speeds of 1,160 or 11,800 RPM, they are connected to a gear reducer. Reducers have two or more gears that work together to reduce the rotational speed of the motor and increase the torque produced. They typically come in standard ratios such as 10:1 or 50:1. This means that if the motor is running at 1,800 RPM, the output shaft of the reducer is running at 1/50 of that, or 36 RPM. Figure 1.2 shows some common reducer configurations. Working from the left and going counterclockwise, there is a right-angle reducer. The right-angle reducer can also be referred to as a worm gear, helical worm gear, or helical bevel, depending on the type of gearing inside. The other two are both parallel or inline reducers because the input and output shafts are parallel to each other. these can also be referred to as cycloidal, planetary, helical, or spur gear depending on the type of gearing inside. Note that in this particular illustration, one of the reducers does not have an output shaft. In these cases, the reducer is mounted directly on the shaft it is driving. This is known as a Shaft-Mounted reducer.

FIGURE 1.2 Gear reducers.

TABLE 1.2
Motor & Reducer versus Gearmotors

	Motor & Reducer Combination	Gearmotor
Ratios	• Limited ratios mean that conveyor speed choices are limited for direct drive applications • Fewer ratios mean fewer spare parts to stock	• There are many more ratios to choose from so it is easier to find one that provides the correct speed for direct drive applications
Efficiency	• Drive efficiency is about 85%	• Drive efficiency can be as high as 95%
Separate units vs single unit	• Being separate units, if the motor or reducer fails, only that unit needs to be replaced. • Being separate units, the overall drive size is larger.	• Being a single unit, if either the motor or reducer fails, the entire unit needs to be replaced. • Being a single unit, the overall drive size is smaller, more compact.

Additionally, notice that the right-angle reducer has a flange to accept a C-Face motor as does the lower Inline reducer. The right-angle reducer has a very shallow flange because the output shaft of the motor fits right into the reducer's input shaft. The Inline reducer has an elongated flange because it uses a three-piece coupling to tie the motor shaft to the reducer's input shaft. Keep in mind that the reducer's flange must be the same mounting designation as the motor that mounts to it. Those designations, such as 56C or 182TC, refer to a series of standards that dictate the diameter of the mounting flange, the mounting bolt size, and the motor's shaft diameter.

Lastly, let us look at gearmotors. Gearmotors are basically a gear reducer and motor together in an integrated unit. As you can see in Figure 1.3, they come in configurations very similar to the reducers discussed earlier. And like those reducers are available with different types of gearing. Gearmotors are referred to by their output RPM rather than ratio or motor RPM.

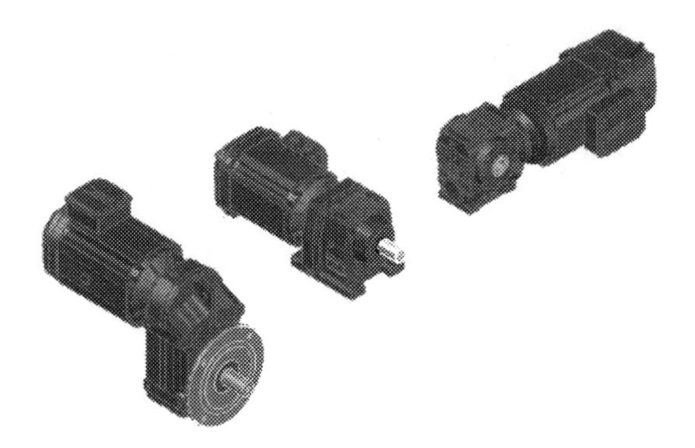

FIGURE 1.3 Gearmotors.

The mounting of the reducer or gearmotor plays a big role in the design of the conveyor drive. For the sake of convenience, in this discussion, when we discuss a reducer, it can also apply to a gearmotor. For instance, a reducer with an output shaft requires either a chain and sprockets, v-belt and sheaves, a timing belt and sheaves, or a coupling to connect it to the drive shaft of the conveyor. Anytime the reducer is directly coupled to the drive shaft of the conveyor, either by a physical coupling or by the reducer being mounted directly on the drive shaft, the conveyor drive shaft will run at the output speed to the reducer.

Keep in mind that anytime more mechanical devices are added to a drive train, the mechanical efficiency will suffer. While a motor and reducer combination may have a mechanical efficiency of 85 percent by themselves, if you add a chain and sprockets, you will lose another 5 percent (see Table 1.1).

When we talk about efficiency, it doesn't seem like a big deal when the drive is 85 percent efficient versus 95 percent until you accumulate the impact of many drives over an entire conveyor system. Then this efficiency can have a real impact on the electrical power consumption of the system.

There are advantages and disadvantages between a motor/reducer combination versus a gearmotor (see Table 1.2).

1.5 KEEPING IT MOVING

The one component that virtually every conveyor has is bearings. Bearings are used to keep shafts turning freely. Figure 1.4 shows five of the most common types of mounted bearings. Starting on the left and going clockwise, there is the four-bolt flange bearing, two-bolt flange bearing, stamped steel pillow block bearing, take-up bearing, and a pillow block bearing. Beyond the housing for the bearing, there are several other differentiating factors in bearings. One is the shape of the bearing between the inner and outer races, such as ball bearing, roller bearing, and spherical

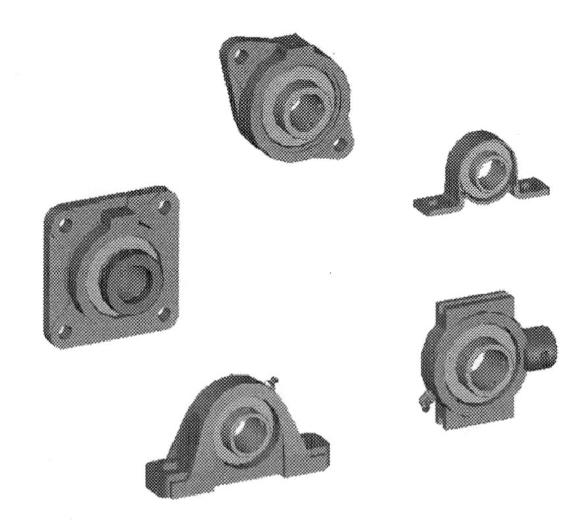

FIGURE 1.4 Bearings.

roller bearing. The bearings without grease fittings are lubed-for-life with the others are re-greaseable. Also important is the mechanism used to lock the inner race to the shaft that passes through it; set screw collar or eccentric locking collar. Then there are the types of seals that keep dirt and debris out and whether the bearing has shields to protect the seals.

1.6 SUPPORTING IT ALL

All conveyor equipment is either supported from the floor or hung from the ceiling. There is a myriad of support types used, but some basic guidelines should be followed.

When floor supports are used, adjustability is required. Typically, floor supports are manufactured to a specific height with adjustment for a minimum of plus or minus 75 mm (3 in.). This allows for floors that are uneven or not level.

If conveyor supports are going to support heavy loads or be subjected to high vibration, once the conveyors are installed at the proper elevation, the supports should be welded to prevent slippage during operation. Slippage can damage the conveyor as well as the product being conveyed, and it can pose a serious hazard to personnel if a conveyor support suddenly slips and product topples.

Whether the conveyor is supported from the floor or hung from the ceiling, diagonal bracing is important to eliminate lateral movement of the conveyors. In the case of a simple straight conveyor with no curves, it is very easy for the entire conveyor to simply pivot off the supports and collapse. Diagonal bracing will eliminate that risk.

It falls to the customer to ensure that the building is strong enough to support the added weight of the conveyor system. The floor must be strong enough to support the additional load and the roof structure must be strong enough to support any ceiling-hung load. Enlisting the services of a structural engineer or an architect is strongly recommended. Typically, your conveyor supplier can give you the point loading for your conveyor system based on fully loaded conveyors.

A liquor distribution system I designed had eight accumulation conveyors in a central location. Working with the architect, we were able to increase the number and size of bar joists in the roof structure to support the conveyors, maintenance catwalk, and accumulated product.

Yet another system had so many conveyors to be hung that the roof structure could not support it. Instead, a mezzanine was installed, and the conveyors were supported from that. The mezzanine loading was calculated, and it was determined that additional footings had to be poured on which the mezzanine would sit. In this case, reinforcing the floor was far less expensive than increasing the structural support of the ceiling, but the cost of the mezzanine tipped the scale the other way. The mezzanine, however, provided other benefits, such as easier maintenance access and additional storage for other supplies. The customer weighed the options and cost and made the determination to install a mezzanine. Each case is different and must be approached from a cost–benefit point of view.

Many times, the system is being installed in a leased building, and there is no opportunity to support overhead conveyor from the roof. In those cases, it requires some creativity. Pallet rack is frequently used to support overhead conveyors. Also, fabricated columns and bar joists can be used to hang the conveyors from. The bar joist is especially useful when trying to span longer distances.

Many municipalities require Professional Engineer (PE) stamped drawings of any structures that will be supporting conveyors that are over 30″ tall. This is especially true in seismic areas. This can add to the cost of the project as well as add time to the schedule.

1.7 RULES OF THE GAME

An organization named CEMA, the Conveyor Equipment Manufacturers Association, and its member companies are the leading conveyor and conveyor system providers. They have written many "standards" for most types of conveyors, such as screw conveyors and belt-driven live rollers conveyors. They also have standards on pulley design and placement of safety stickers and signs. Some manufacturers adhere strictly to CEMA standards; others will bend or even break those rules outright.

The key to breaking the rules is understanding the consequences. CEMA's standards are intended to provide the most robust, long-running conveyors. Sometimes, and this should be the exception, bending or breaking the rules of conveyor design will solve a problem, but there is a cost associated with varying from the rules. There will be case studies in various chapters discussing where the rules were broken and the associated costs. NEVER compromise safety to solve a problem! Safety rules are never to be bent or broken.

1.8 QUESTIONS

1. List two objectives of any conveyor application.
2. Which of these are *not* a part of the cost of a material handling system?
   a. Periodic maintenance
   b. Spare parts
   c. Enterprise resource planning (ERP) software
   d. Cost of stocking spare parts on site
3. Which of these is not an objective of a material handling system?
   a. Better record keeping
   b. Improve workflow
   c. Perform all handling operations at the lowest reasonable cost
   d. All of the above
4. Which drive design is more efficient: motor and reducer combination or gearmotor?
5. Which drive design requires the fewest spare parts?
6. What is the advantage of lubed-for-life bearings?

2 Equipment Selection Guide

Nothing replaces experience, but the information that follows will help you ask intelligent questions and make intelligent decisions. An old hand who has kept current with the ever-changing picture in the material handling field can draw on his or her knowledge and make a specific selection quickly. Even novices can end up with the correct selection, if they tackle the problem in an organized sequence of steps.

First, make the equipment selection carefully and honestly, by analyzing the problem. Second, before settling on a solution, be sure you have reviewed every product that is to be handled and have considered all operational features against the productivity to be gained, as well as the cost of the equipment against the earnings to be gained through the use of that equipment.

2.1 CONVEYOR SELECTION

Follow these simple steps to select the conveyor most appropriate for your application:

1. Identify the product(s) to be conveyed.
2. Use the following list to identify which selection chart(s) to use. Charts can be found at the end of this chapter.

Baggage	Chart I
Bags	Chart I
Bundles and bales	Chart I
Cans and bottles	Chart A
Cans, drums, and pails	Chart H
Corrugated cartons	Chart C
Crates	Chart C
Loose, bulk material	Chart B
Metal pallets	Chart E
Plastic pallets	Chart G
Slip sheets	Chart F
Tires	Chart H
Totes	Chart C
Wheels	Chart H
Wire baskets	Chart G
Wooden pallets	Chart D

3. Review the selection notes in the next section, and then refer to the applicable chart(s).

DOI: 10.1201/9781003376613-2

2.2 NOTES

2.2.1 Corrugated Boxes

- Good condition: Taped or sealed cartons with dry, flat, and firm conveying surfaces and of sufficient strength to support their own weight.
- Fair condition: Dry, sealed, and unsealed cartons with bent edges or corners and minor nonflat or bulging conditions.
- Poor condition: Battered cartons that are apparently soft due to dampness or wetness, or that have irregular bottom surface and uneven weight distribution, and barely have sufficient strength to support their own weight.

2.2.2 Straps, Binding, etc. on Boxes

To ensure optimum conveying of all products, they must be free from protruding or loose straps, hardware, and corner reinforcements. Wire binding and steel straps on cases must be relatively flush with the carrying surface if a roller or wheel conveyor is to be used.

2.2.3 Pallets

Bottom boards on a pallet must be strong enough to support the load without deflecting significantly.

2.2.4 Unusual Products

Many products just do not convey as we expect them to. Some of the more challenging products are five-gallon buckets, drums, wheel rims, or plastic pallets. The challenges arise from the shape of their bottom surface. A bucket or drum has what is known as a chimed bottom, which refers to a container that is supported by a rim or flange around the outside edge. The same holds true for pallets that have bottom boards only around the outside periphery.

These products will typically travel properly on roller conveyors. Challenges occur when a transfer is needed or if the product is conveyed on a chain conveyor or other narrow conveying surface. When the product reaches the end of the conveyor, it will need additional support while it transfers onto the next unit because the product is only supported on the outside rim. The chimed bottom will drop off the end unless there is additional support for the outer edges. The other alternative is to have some overlap in the conveyors so that the head or drive end on one conveyor actually goes past the tail of the next conveyor. The overlap ensures that the product is never unsupported. There are several solutions to this issue; each manufacturer will have its own approach, and each application will have its own requirements. It is important that you are aware of these challenges.

If a specific product has any characteristics that might make it difficult to convey, test the product on the selected conveyor to confirm suitability.

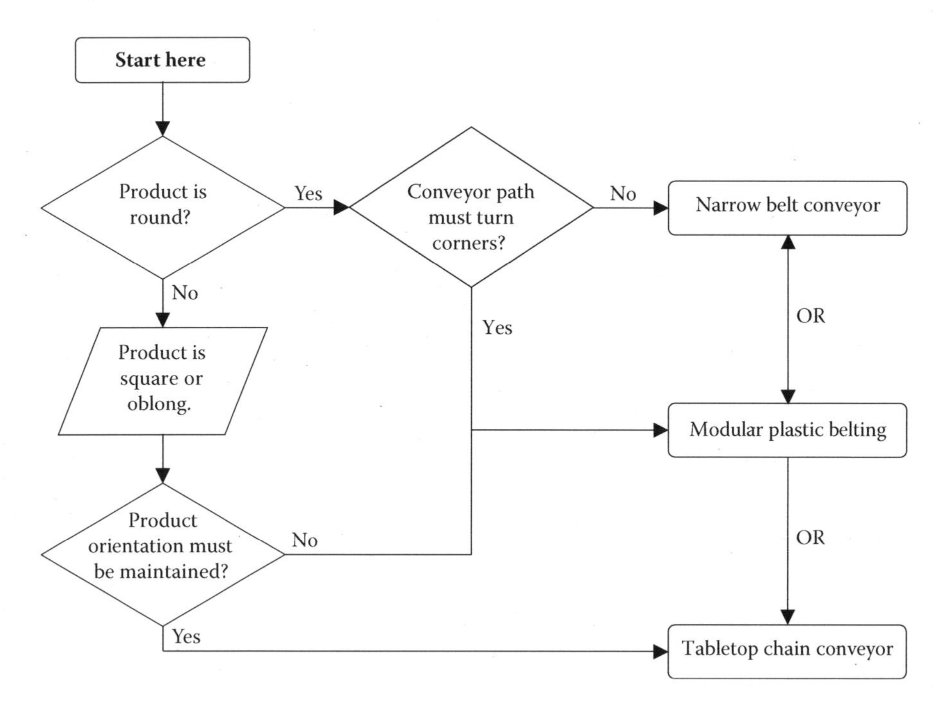

CHART A Bottles and cans.

Always keep the application in mind and think about how different types of conveyors can be utilized. As an example, when conveying disposable cameras, a tabletop chain conveyor would be a natural choice to carry the product between process machinery. The better choice would actually be narrow, two-inch-wide belt conveyors using a needle-stitched belt. The needle-stitched belt has a felt-like conveying surface so it would not scuff the bottom of the cameras, whereas a tabletop chain could.

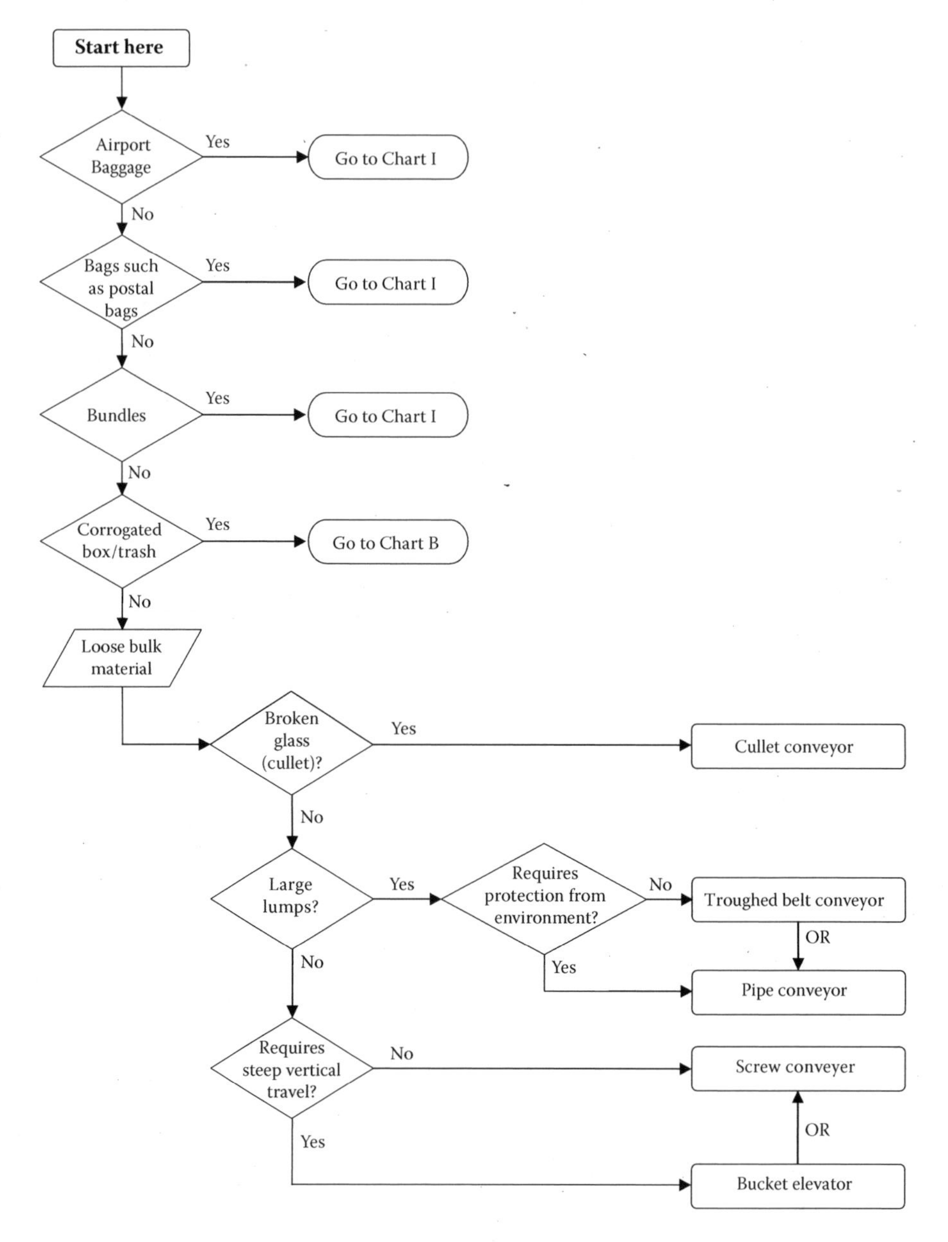

CHART B Loose, bulk material.

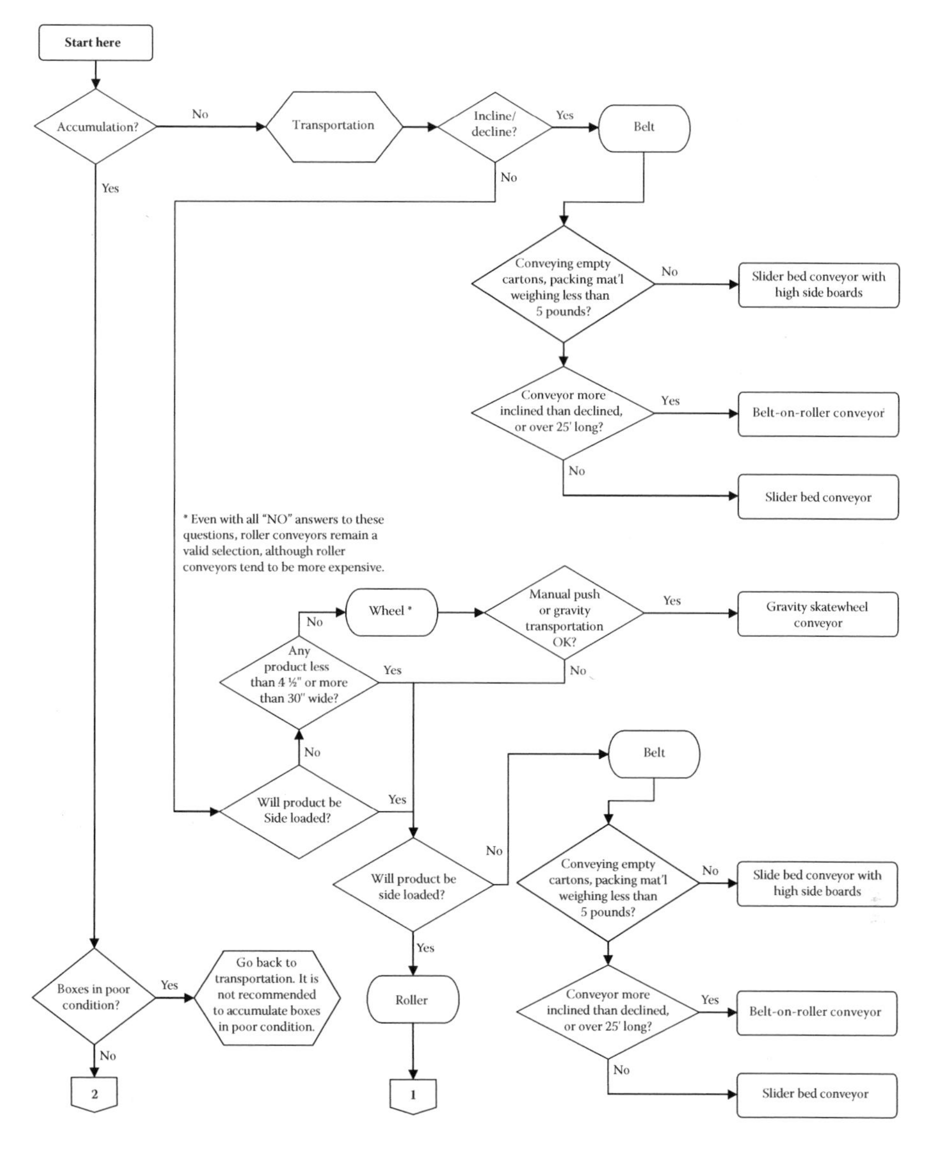

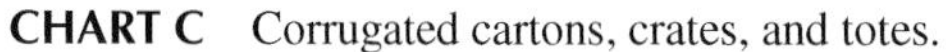

CHART C Corrugated cartons, crates, and totes.

(Continued)

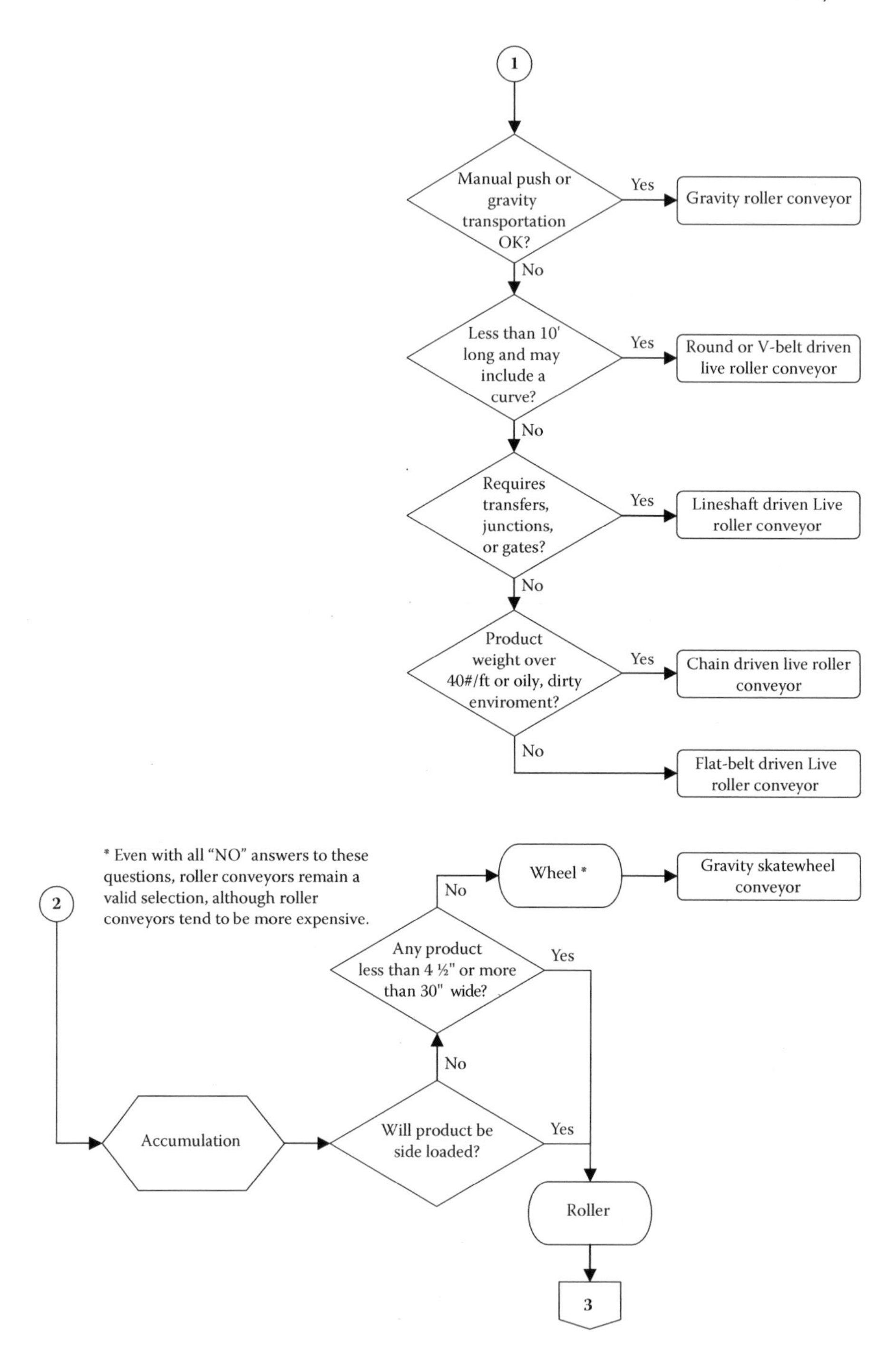

CHART C (Continued) Corrugated cartons, crates, and totes.

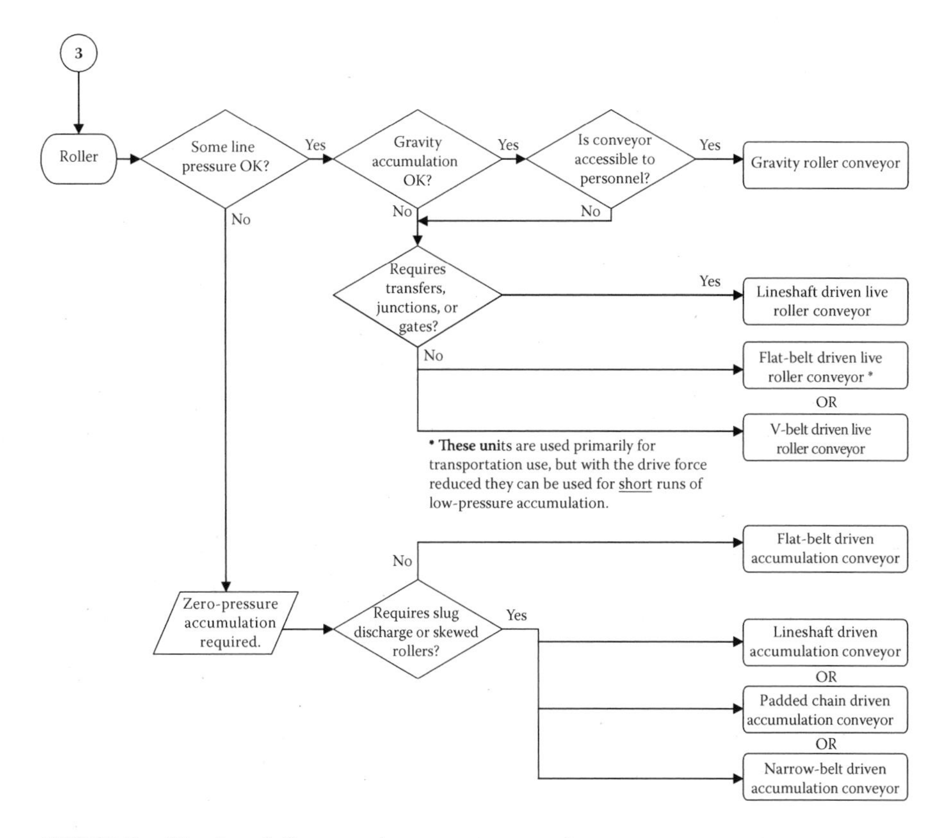

CHART C (Continued) Corrugated cartons, crates, and totes.

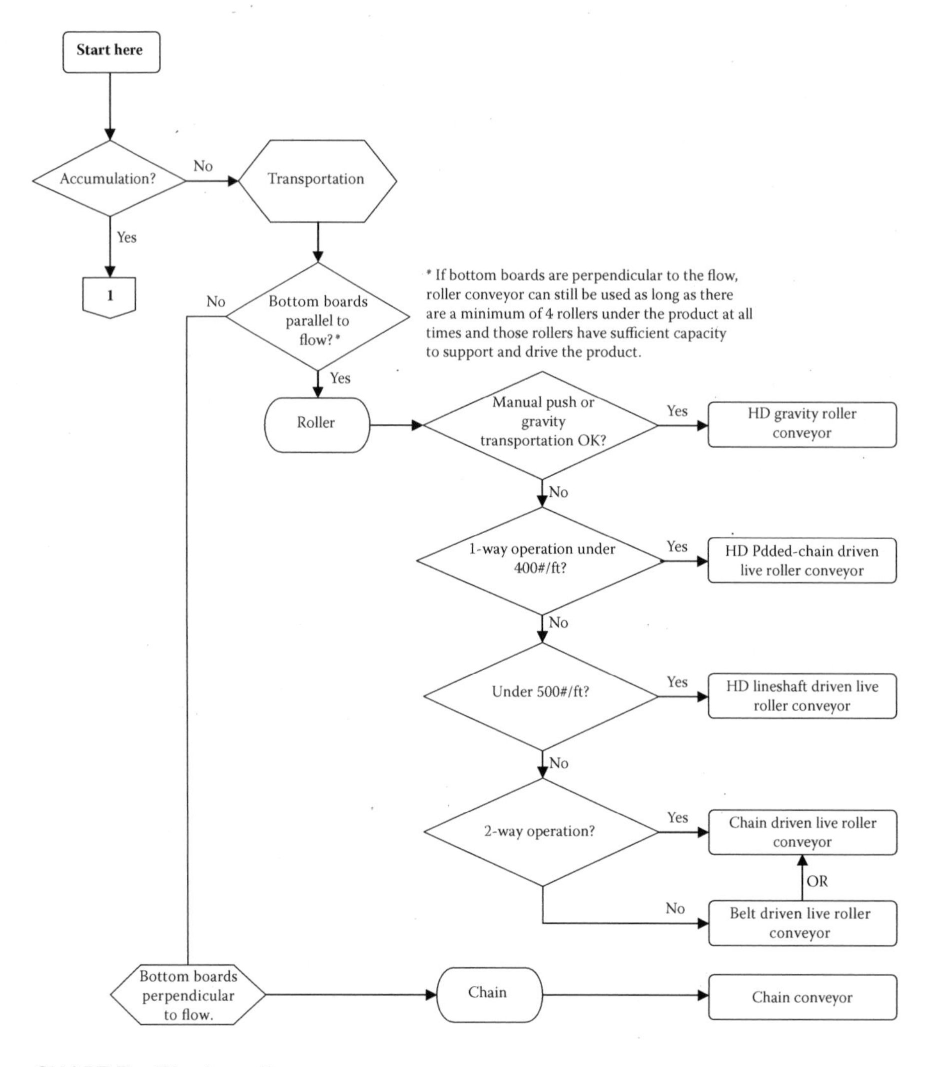

CHART D Wooden pallets.

(*Continued*)

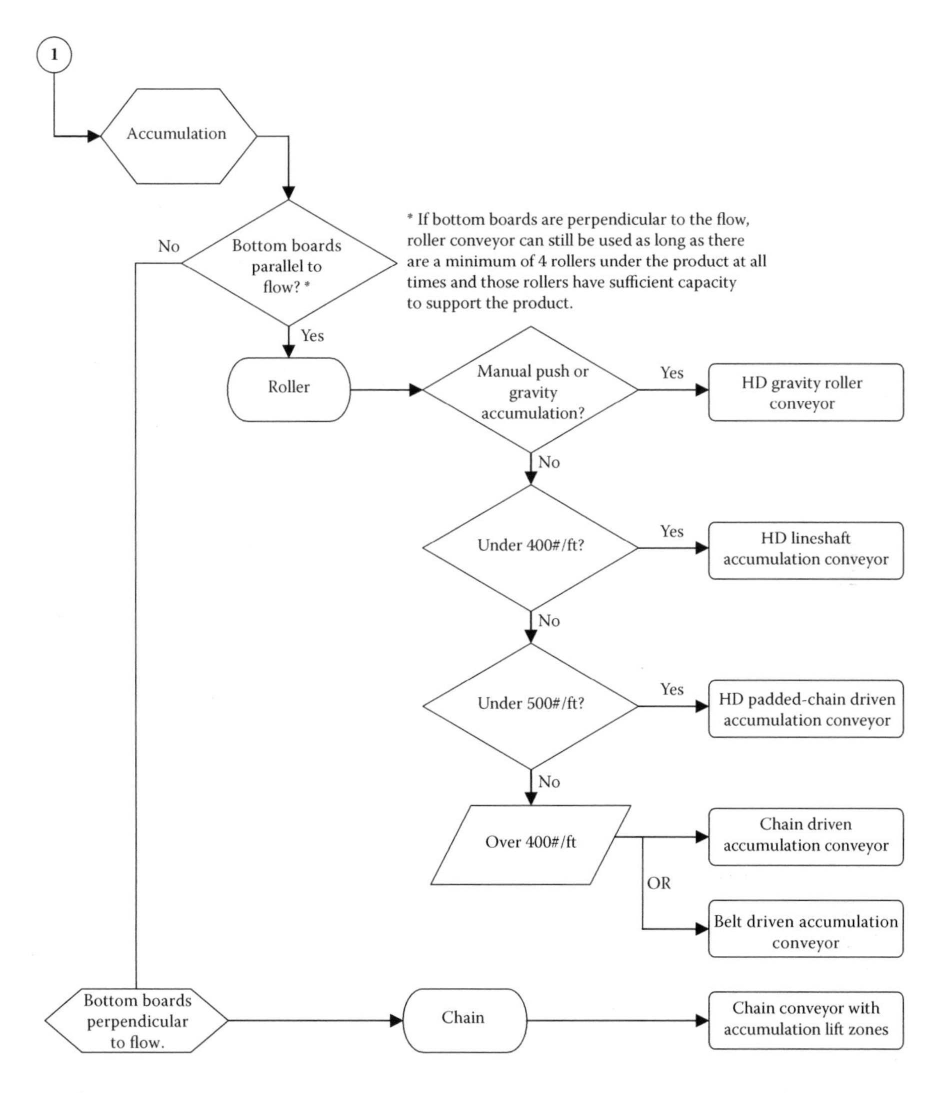

CHART D (Continued) Wooden pallets.

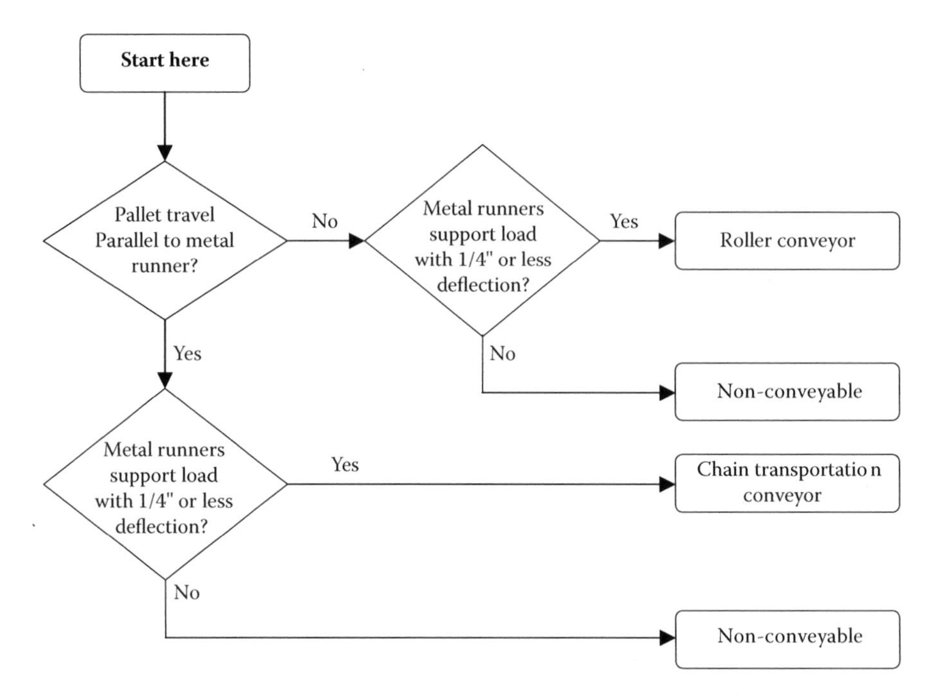

CHART E Metal pallets.

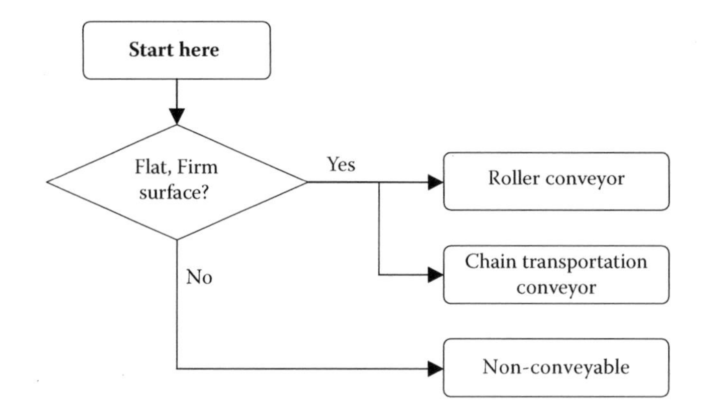

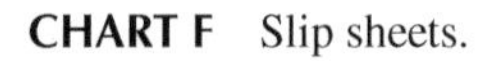
CHART F Slip sheets.

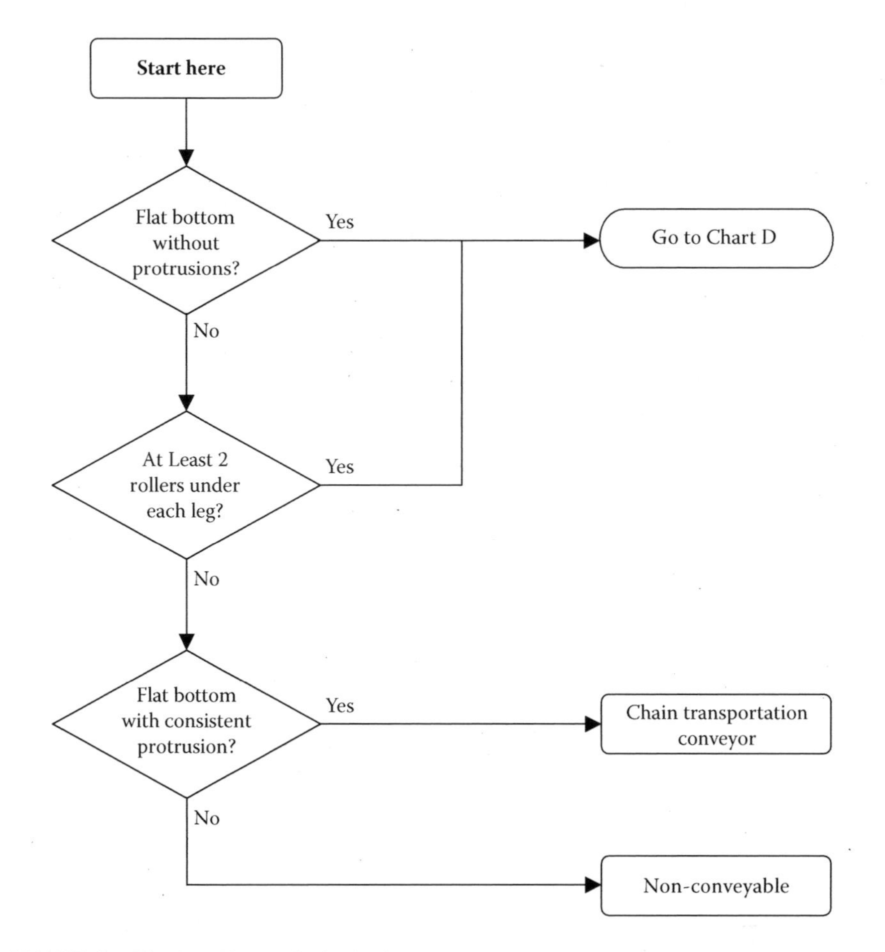

CHART G Plastic pallets and wire baskets.

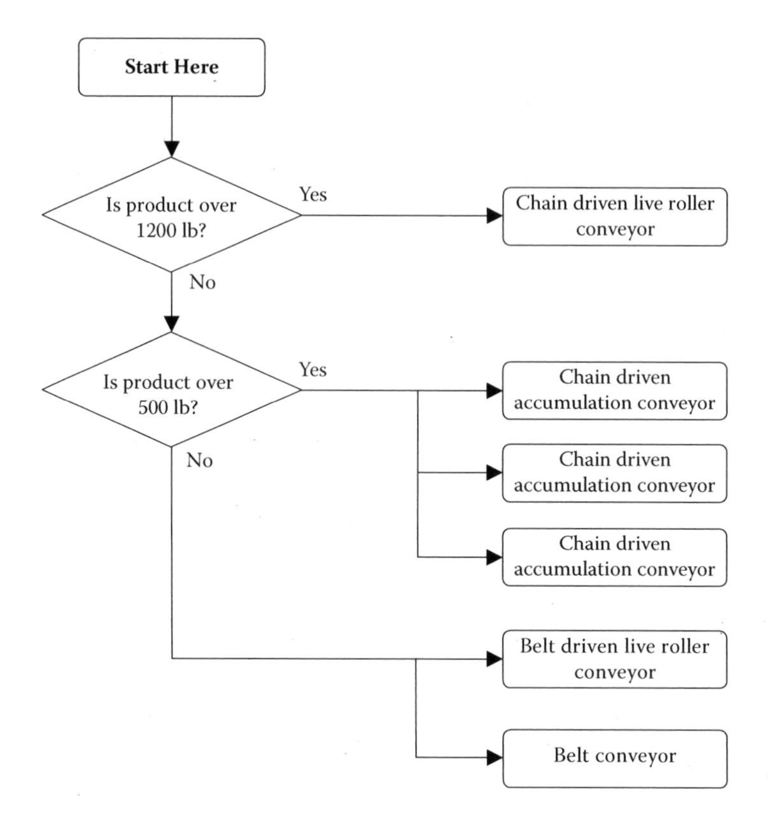

CHART H Cans, drums, pails, tires, and wheels.

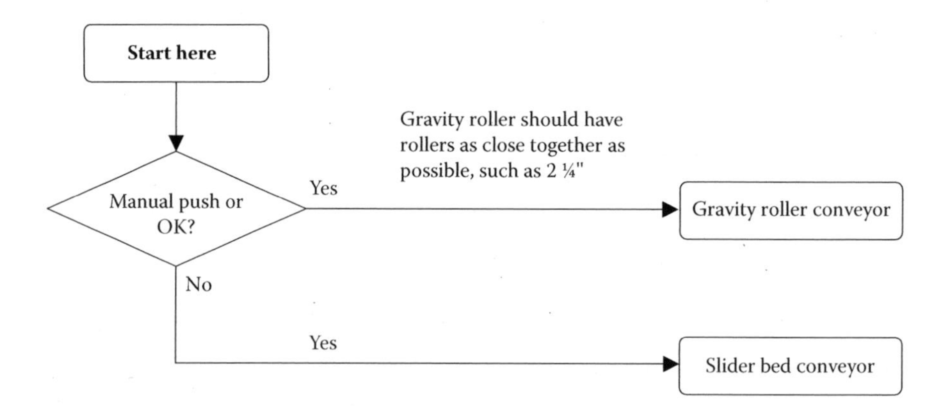

CHART I Baggage, bags, bundles, and bales.

2.3 QUESTIONS

1. What type of conveyor would you select for carrying empty round glass bottles and why?
2. If you needed to accumulate a product prior to a piece of machinery that was 24″ wide x 36″ long x 12″ high and weighed 200 pounds, what conveyor would you select and why?
3. What type of conveyor would you select for carrying a pallet of steel drums with chimed bottoms that weighed a total of 2,700 pounds (1,225 kg) and why?
4. What type of conveyor would you select for conveyor plastic 55-gallon drums that are to be filled with oil and that are 24.5″ (622 mm) in diameter and weigh 900 pounds (400 kg) when filled?
5. If you are going to select a conveyor to handle empty maple syrup bottles, which of the following questions would you ask?
   a. Is the product round or something else?
   b. Does product orientation matter?
   c. Does the product need to accumulate?
   d. All of the above.
6. You have been tasked with determining the type of conveyor to handle sand and transport it a mile in distance. Which of the questions below would apply?
   a. What kind of sand is it: frac sand, masonry sand, etc.?
   b. Should the product be protected from the wind or rain?
   c. What is the sand used for?
   d. Is the sand recycled?
7. Handling bags of dog food or kitty litter can be challenging due to their flexibility. What is the best conveyor type for handling individual bags?
8. What type of conveyor would be a good choice for handling rectangular motor oil bottles prior to going into a labeler?
9. You have been tasked with finding a conveyor to handle a variety of different-sized plastic pallets, but none of them weigh over 500 pounds. Some have six legs, some have nine legs, and some have a solid uniform bottom. What type of conveyor would you choose to make the most types of pallets conveyable?
10. If you were to develop a carton handling system that needs to divert packages to various other areas and also have gates that allow people to access those areas?

3 Tabletop Chain Conveyor

3.1 TABLETOP CHAIN

There are three primary types of TTC conveyors. The first is when the top plate is an integral part of the link, as shown in Figure 3.1a. The second is when the top plate is snapped onto a base chain, as shown in Figure 3.1b. The third is referred to as multiflex or EWAB (EWAB is the name of the company that originally developed that chain in 1970), and it utilizes a ball joint to provide a greater amount of flexibility, as shown in Figure 3.1c.

All three types are available in special purpose configurations. For example, Figure 3.2a shows how rollers can be added to the top of the chain to use it for low-pressure accumulation. A rubber insert can be added to use the chain to carry product up an incline, as shown in Figure 3.2b. Gripper attachments can be used when two face-to-face chains are used to grip a product to carry it without any support underneath (see Figure 3.2c).

There are many unique chain top plate configurations available. The typical top plate can be redesigned to minimize gaps to reduce possible vibration when the product is stopped on the chain as shown in Figure 3.2d. Also, Figure 3.2d shows they can have vertical ribs to minimize contact with the product, thus reducing surface friction. The top plate can be perforated to provide airflow to either blow air through to cool the product or suck air through with a vacuum to more positively hold the product in place, as shown in Figure 3.2e. Figure 3.2f shows cleats can be added to aid in moving the product or to provide separation for the product. There are a myriad of configurations available based on the product to be handled.

TTC conveyors are frequently used for plastic and glass bottling plants. Originally, TTC was only available in a straight chain. This required the use of deadplates or transfer disks to change directions. The raised rib chain shown in Figure 3.2(d) allows for the use of fingered deadplates for a smoother transition. A side-flexing chain has greatly improved the versatility of TTC. Side-flexing chains also allowed conveyor systems to run faster because deadplate transfers were a serious source of product instability and greatly reduced the speeds at which product could be conveyed. TTC conveyors are traditionally available in widths ranging from 3¼ inches to 12 inches. TTC is available in a variety of plastics (e.g., acetyl or Delrin®), steel, or stainless steel. Some manufacturers offer plastic formulations that include Microban® to limit bacterial growth.

The typical configuration for a TTC conveyor frame is a formed steel or stainless-steel frame with wearstrips on top to guide the TTC, a support for the return chain, and guiderails to keep the product on the conveyor. Some manufacturers offer extruded aluminum conveyor frames. Figure 3.3a shows a typical open frame design and Figure 3.3b shows a common closed channel design. The open frame design is

DOI: 10.1201/9781003376613-3

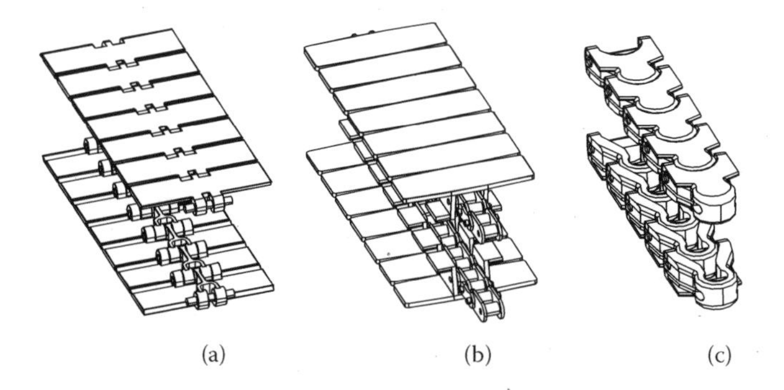

FIGURE 3.1 Tabletop chain: (a) straight-running, (b) side-flexing, and (c) multiflex.

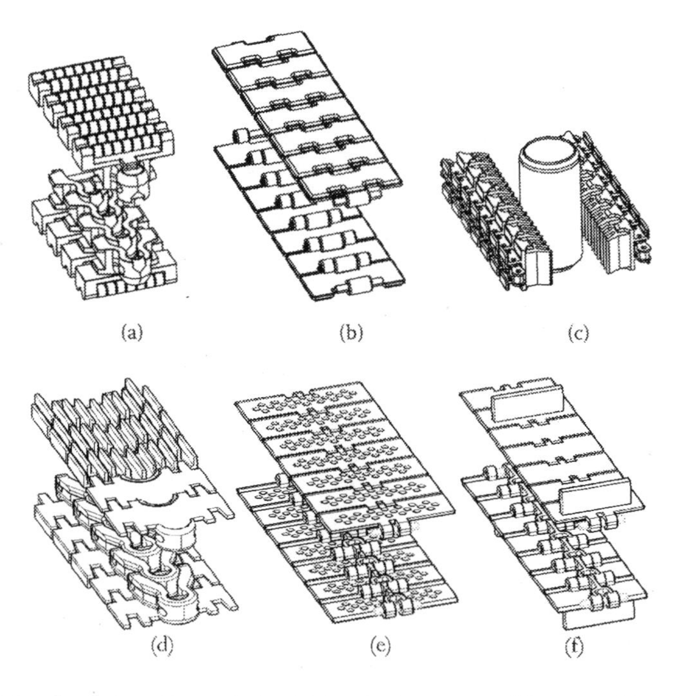

FIGURE 3.2 Specialty tabletop chain: (a) low back pressure, (b) non-skid top, (c) gripper chain, (d) raised rib, (e) perforated, and (f) cleated.

popular in washdown applications because horizontal surfaces to retain any debris are minimized. They also allow water to pass through the frame easily. The carry wearstrip is most frequently made of ultra-high-molecular-weight polyethylene (UHMW) either over an aluminum extrusion or around the edge of the steel frame. Other low-friction materials are used, such as Arguto (a wax- and oil-impregnated wood) that is used primarily for steel and stainless-steel chains and cold-rolled steel on a wood spacer used primarily for plastic chains. The three most common ways the return chain can be supported are a pan with a low-friction surface (UHMW or Formica), rollers, and UHMW (straight extrusions or serpentine).

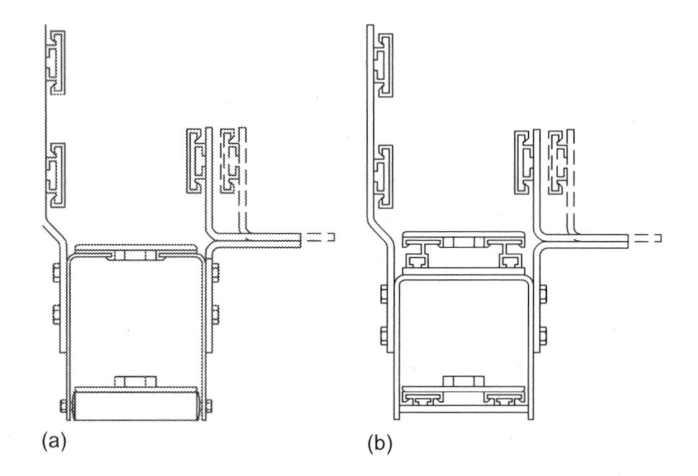

FIGURE 3.3 (a) Open frame design with return roller and (b) closed frame design with UHMW return.

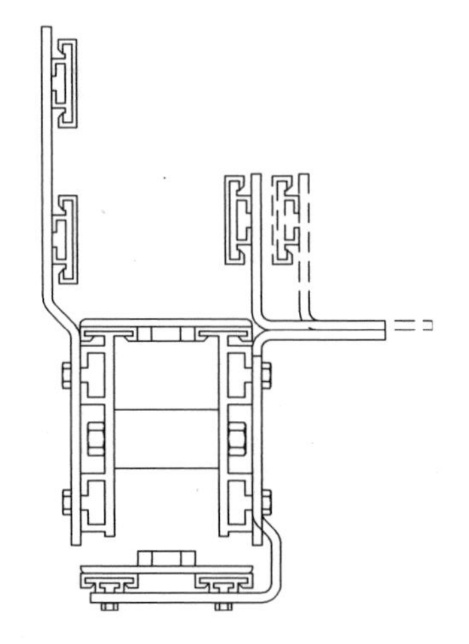

FIGURE 3.4 Extruded aluminum design.

The third configuration, shown in Figure 3.4, uses an extruded aluminum frame. This is just one of the many variations that are available.

As a system is developed, a conveyor might become too long. In this situation, it is necessary to transfer the product to a second conveyor. This is most commonly accomplished with a parallel or side transfer. Figure 3.5 shows a parallel transfer. In the use of TTC conveyors, the smooth transfer of the conveyed product from one chain to another is essential for product stability, protection, and reduced system downtime due to jams.

Parallel transfers are the least expensive and preferred method of making a product transfer. Although simple in theory, care must be taken to assure that the chain strands are operating at level with each other or with the outfeed chain slightly lower.

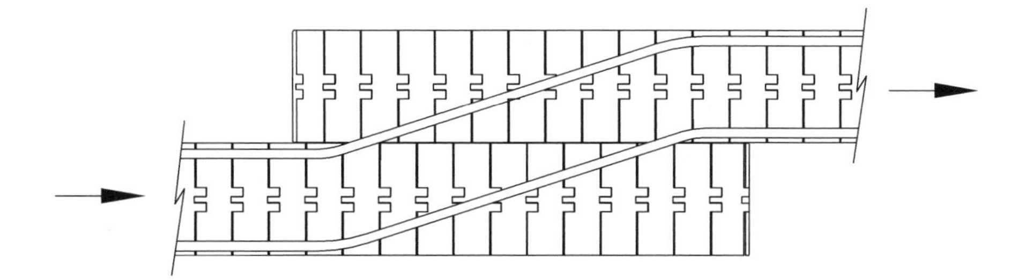

FIGURE 3.5 Parallel transfer.

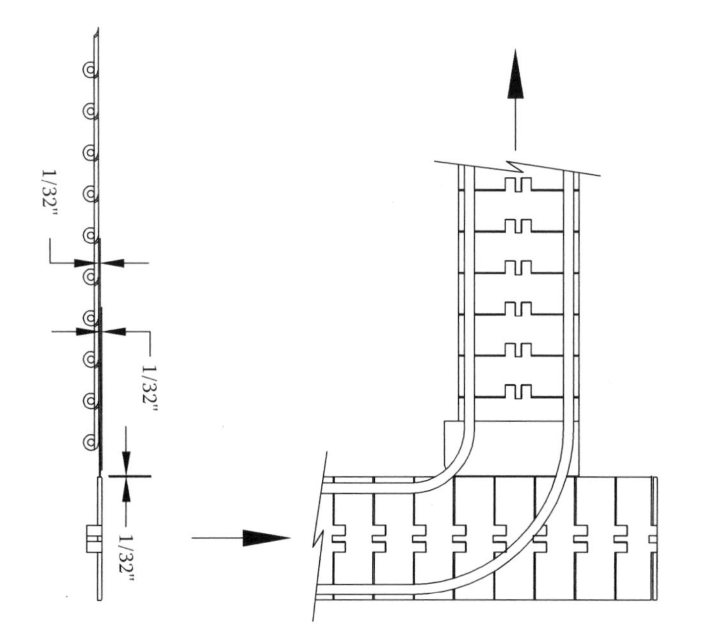

FIGURE 3.6 Deadplate transfer.

The gap between chains should be minimized, the guiderails need to be adjusted to provide a smooth transfer, and the speeds should be carefully considered.

When TTC was first introduced, it could only run in a straight line. When it became necessary to turn a corner, the product had to be transferred to another conveyor. This required special strategies for affecting a smooth transfer. There are two primary methods of changing direction: a deadplate transfer (see Figure 3.6) and a turntable transfer (see Figure 3.7).

A typical deadplate transfer is shown in Figure 3.6. For the smoothest operation, the deadplate should be mounted so that it is perfectly level with or slightly higher than the outfeed chain at the highest chordal position of the sprocket. The deadplate should have a slight bevel to the end and be as thin as possible so that the product is conveyed onto and off of the deadplate without disruption.

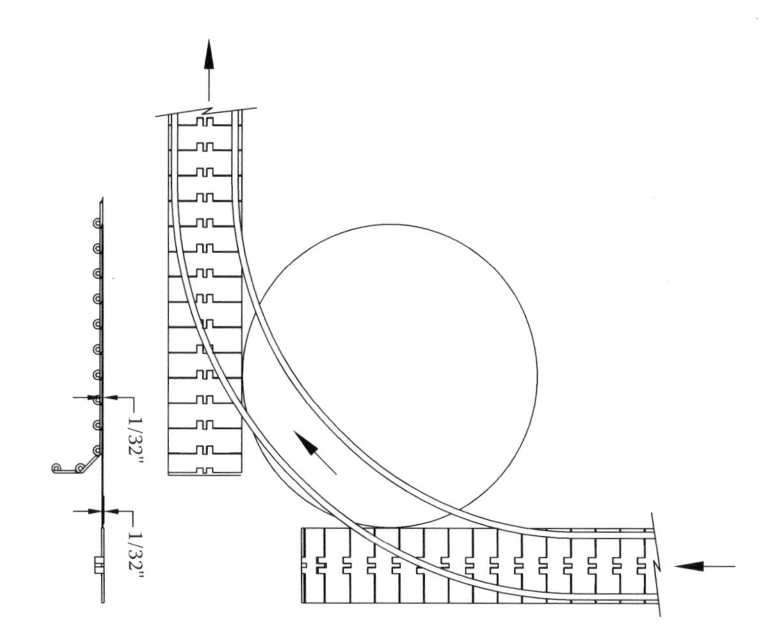

FIGURE 3.7 Disk transfer.

The principles behind turntable or disk transfers are basically the same (see Figure 3.7). Alignment is critical in assuring smooth product transfer. The edge of the turntable is typically beveled so that there is no sharp edge on which the product can get caught. Turntables should be mounted slightly lower than the infeed conveyor and slightly higher than the outfeed conveyor. The rotary motion of the turntable is typically provided by a small right-angle gearbox slave driven from the tail sprocket of the outfeed conveyor.

TTC is available as both straight-running and side-flexing chains. The side-flexing chain is more expensive, so it is usually used only where needed. There are two primary classifications of side-flexing TTC: bevel and tab. These terms refer to how the bottom of the chain engages the wearstrip through the curves. As illustrated in Figure 3.8, bevel chain uses a dovetail-style engagement to hold the chain down through the curves, whereas a tab chain uses the more positive tab actually riding under the wearstrip to hold it down. Using a bevel chain, the straight sections do not have to engage the bevel.

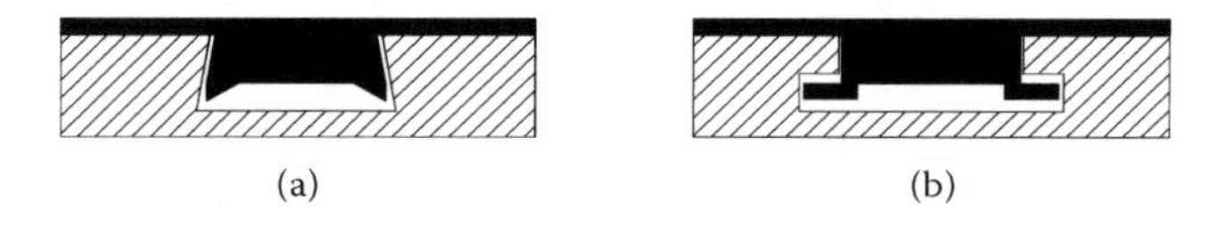

FIGURE 3.8 Curved chain and wearstrip cross-sections: (a) bevel style chain and (b) tab style chain.

TTC curves have limitations on the minimum radius of the curve. Typically, most TTC chains (4½-in. up to 7½-in. wide) have a minimum centerline radius of 24 in. Chains 10-in. and 12-in. wide have a minimum centerline radius of 30 in. Chains 3¼-in. wide typically have an 18-in. minimum centerline radius, as do the roller-chain-based chains.

When conveying small or unstable product on a side-flexing chain, there is one frequently overlooked aspect. Straight-running chain links are rectangular and have a minimal gap between links across the entire width of the link. The links of side-flexing chain, on the other hand, are tapered to the outside to allow clearance for turning. The wider the chain, the wider the gap between adjoining links at the outside of the curve. As the chain goes through a curve, the gap toward the inside of the curve closes and the gap toward the outside opens further. This gap can cause small or unstable product to fall over. In these cases, it is better to use two or three narrower chains rather than one wide chain. For instance, instead of one 10-in-wide chain, use two 4½-in. chains or three 3¼-in. chains.

TTC curves, just like the straights, come in a variety of constructions. Besides the frame design, there are a variety of wearstrip designs for curves. The wearstrip can be a series of machined sections of Ceram-P, a glass bead-filled UHMW that is typically green in color. Some manufacturers use precision-ground, hardened steel wearstrips. Commercially available wearstrips are also machined from solid plastic. Depending on the manufacturer, the material used is UHMW, Nylatron, Ceram-P®, or Delrin. Because they come premade to required tolerances, they offer a simple solution when trying to build a curve quickly without performing a lot of highly accurate machining or assembly.

3.1.1 Drive Requirements

A drive is required to make the chain move. In virtually all cases, TTC conveyors use an end drive. At the drive end of the conveyor, directly below and behind the drive sprocket, there is a sagging loop of chain called the catenary. This catenary loop is used to control the chain's tensions and is generally 18-in. long and 3- to 5-in. deep. The drive itself can be mounted in a variety of manners. The most popular are shaft-mounted drives. Side-mount or underslung drive mounts transmit the power via roller chain or timing belt and matching sprockets.

3.1.2 Guiding the Product

In Figures 3.4 and 3.5, both fixed and adjustable guiderail brackets are shown. There is a wide array of guiderail mounting hardware. Figure 3.9 shows just some of the bracketry that is available. Obviously due to their simplicity, fixed brackets are the least expensive, but they also limit the product that can be run on a conveyor. If an adjustable bracket is used, the guiderails can be readjusted as necessary to accommodate different-sized products. In the cases of very tall product, two or more pieces of guiderail can be used. Ideally, the lower rail will keep a tipped or fallen product on the conveyor and the upper rail should be approximately two-thirds of the way up on the product.

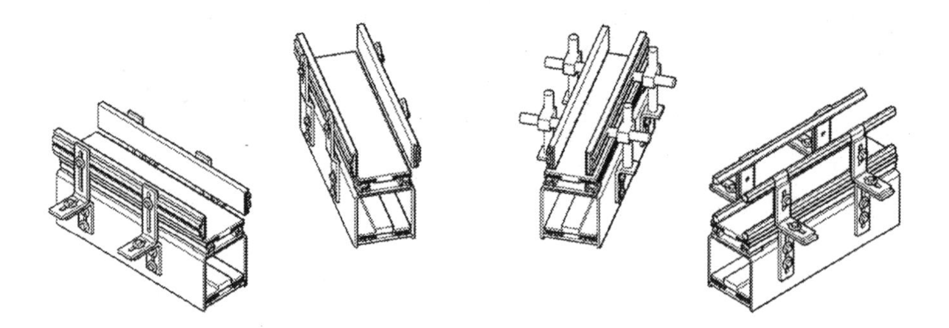

FIGURE 3.9 Sample guiderail brackets.

3.1.3 Multiflex

Much of what we have discussed thus far has been about standard TTC conveyors. However, at the beginning of this chapter, we briefly mentioned multiflex or EWAB chains. Multiflex has the ability to flex in many directions unlike a simple hinged joint as with other TTC. Multiflex conveyors tend to be smaller due to the smaller chain. The conveyor track typically consists of a channel with a serpentine steel wear surface (see Figure 3.10). Antifriction, small-radius turning disks (see Figure 3.11) help reduce chain tension and allow for compact conveyor construction. Typical turning disks allow a centerline radius of only 5.9 in.

FIGURE 3.10 Straight section of multiflex conveyor.

FIGURE 3.11 Multiflex disk turn.

3.2 MODULAR PLASTIC CONVEYOR BELTING

This type of conveyor is similar to TTC in that it uses plastic chains running on wearstrips, but the major difference is the size. Modular plastic belting is frequently used in widths ranging from 3 in. to 12 ft. Modular plastic belting is generally constructed from a series of short, narrow, injection-molded links or modules that, when assembled with long hinge pins, make up the width of the belt. Normally, the belting is assembled in a brick pattern so the joints do not line up from row to row as seen in Figure 3.12a. This holds true for all plastic belts except those that are a single link in width. Intralox invented this technology in 1971.

Most plastic belts are also usable for bidirectional applications, whereas TTC is not.

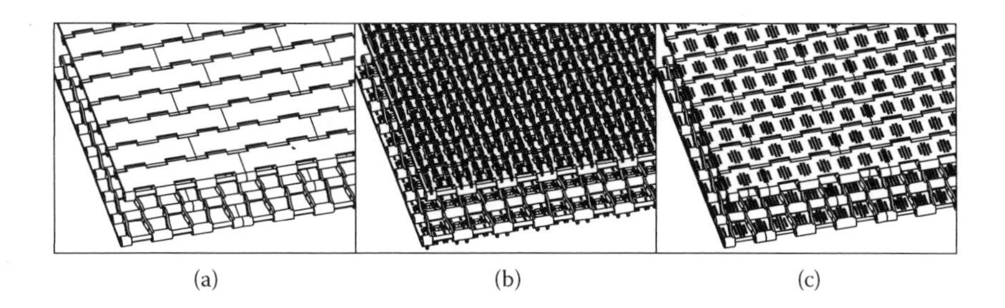

FIGURE 3.12 Plastic belting: (a) flush top, (b) raised rib, and (c) open grid.

Plastic belting is available with a variety of top surfaces:

- Flat top
- Perforated flat top
- Flush grid
- Open grid
- Raised rib
- Friction top
- Roller top
- Nub top (antistick)
- Cone top (extra grip)

Open hinge configurations are used for straight-running and side-flexing applications. Figure 3.12c shows the open grid, which is frequently used for food applications so that it can be washed down and not retain any food particles. Many of these belts carry U.S. Food and Drug Administration (FDA) approvals, which should be verified prior to use. The flush top (Figure 3.12a) offers the smoothest surface for carrying products. This is also offered with various perforations to allow airflow through the belting for vacuum or cooling applications. The raised rib (Figure 3.12b) is used when the product must be transferred off the end of the conveyor, because it allows the use of fingered deadplates rather than straight deadplates. The drawback to the raised rib style is that some products can be unstable due to the limited support surface.

If you look at Figure 3.13, you can see that a straight deadplate can cause product instability due to the chordal action of the belting. Chordal action is the effect of the series of flat plates that bend around a sprocket. Each plate creates a chord on the circumference of the sprocket. Either end of the chord extends further from the center of the sprocket than the center. This creates a conveying surface that moves up and down as the plates flex around the sprocket. Larger sprockets minimize the chordal action but require longer deadplates. Smaller sprockets allow for shorter deadplates, but they make the chordal action more pronounced. This is the same type of deadplate used with TTC as shown earlier in Figure 3.7. The fingered deadplate virtually eliminates that problem.

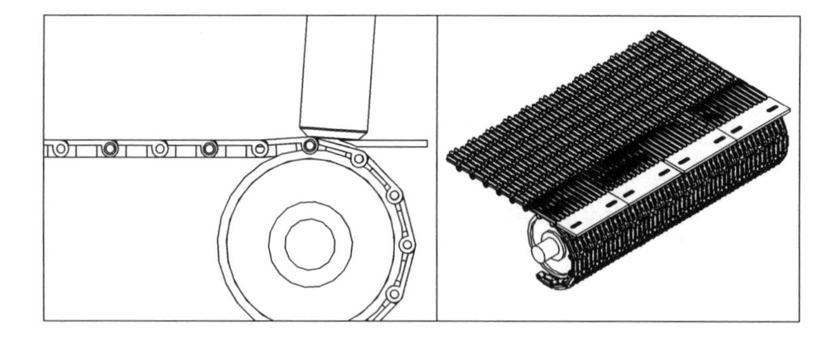

FIGURE 3.13 Transfer detail of a fingered deadplate.

Modular plastic belting uses a series of sprockets on the drive shaft rather than just one. Normally, the center sprocket is locked in place with set screws or clamp collars, and the rest are allowed to float on the shaft to compensate for expansion and contraction of the belt due to heat and humidity.

Unlike typical TTC conveyors, modular plastic belt conveyors do not require an end drive. Because many modular plastic belt conveyors are designed to be bidirectional, center drives are quite normal. Modular plastic belts should not be run in a tight, tensioned manner. As with TTC conveyors, a catenary is required on the slack side of the center drive. If a conveyor is bidirectional, there must be an allowance for a catenary loop on both sides of the drive.

3.3 MOUNTED ROLLER BELT

A newer type of modular plastic belt is called mounted roller belt or MRB, again originally developed by Intralox. This belt has rollers mounted in the belt as shown in Figure 3.14. The key is the direction the rollers are facing as well as the height of the roller above the surface of the surrounding belt. The rollers can roll as the belt moves or they can be actuated separately from the belt causing the product on the belt to move in a direction other than the path of the belt. Some manufacturers offer MRB with balls in the belt rather than rollers. This offers some very unique options when it comes to directing products on the belt by actuating the balls to move the product side to side while the belt is running forward.

In Figures 3.3–3.14, there are four examples of MRB. The top one has the rollers pointed in the same direction as the belt would travel. This allows products to accumulate on the belt with minimal friction. The next one down has the rollers mounted in a transverse direction. This allows products to move left to right on the conveyor

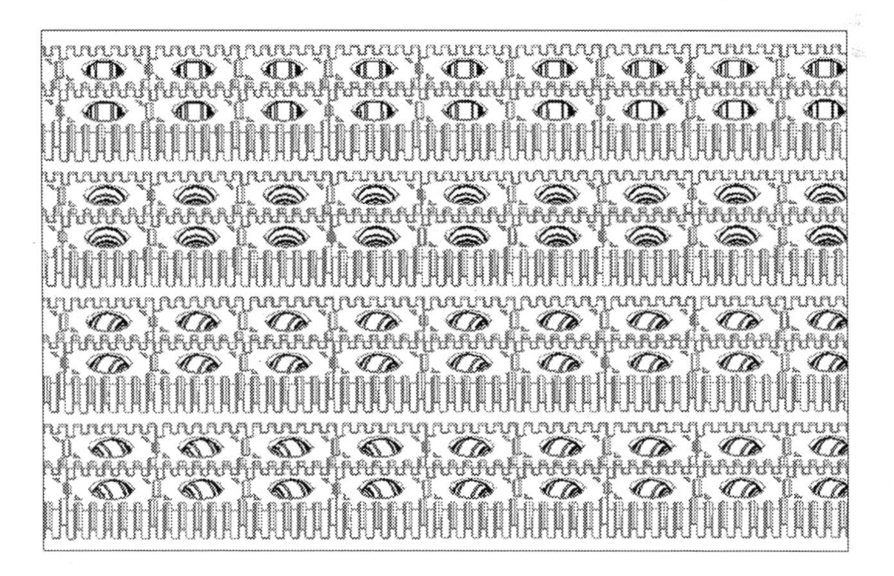

FIGURE 3.14 Mounted roller belt.

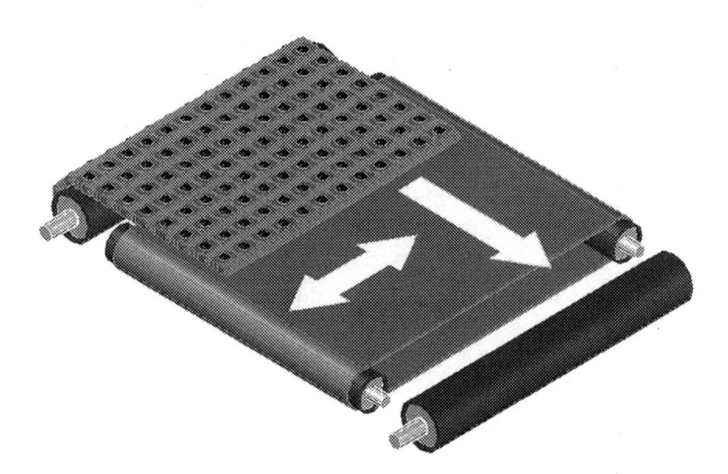

FIGURE 3.15 Mounted roller belt application.

very easily. Next, we have an MRB with the rollers pointed to one side at an angle. This allows products to move in the direction the rollers are pointing as the belt moves. It is very useful in getting product all to one side of the conveyor. Lastly, we have an MRB with rollers at an angle, but half are pointed to the left and half are pointed to the right. This belt would bring all of the products to the center of the belt.

Mounted rollers belts have become very popular in package handling, especially handling small irregularly shaped products such as polybags.

In Figure 3.15, we see one application of an MRB being used as a divert. In this scenario, the MRB has transverse mounted rollers so when the belt underneath the MRB moves to the left any package on top of the MRB would move to the right and vice versa.

3.4 APPLICATION DETAILS

This section goes into more detail on the application considerations of TTC. TTC is used to handle a wide variety of products: glass and plastic bottles, cereal boxes, toilet paper, and almost anything else that must be kept single file. The modular plastic belting is also used for a variety of package handling applications as well. Always keep in mind when specifying a conveyor that whatever is in a container that is being conveyed will be on the conveyor at some point. Whether it is soda, hair perming solution, or maple syrup, it will at some point be spilled on the conveyor and can affect the required horsepower to pull the conveyor. Although many variables are taken into consideration as far as types of lubrication, the effects of maple syrup, for example, on conveyor operation will require some experimentation.

The product being handled will also affect the choice of the wearstrip material used for guiding the chain. One customer I dealt with was manufacturing hair care products. The neutralizer used in hair perming kits was very corrosive. The neutralizer would dissolve the paint off of the conveyor and eat through the steel. The conveyor had to be made from stainless steel and the wearstrip also. The neutralizer would turn the UHMW into a thick spongy material that lost all of its useful low-friction properties.

TTC is most frequently used for products like bottles or cans that need to be conveyed upright for labeling, filling, and so on. When transporting product from one point to another, the product is normally kept in a single file to simplify handling. Any time two products get side by side, they can cause jams and additional work is required to get them in a single file again. One trick is to allow the guiderails to open up to a width of one and two-thirds of the product diameter. This way the product never gets completely side by side so that when the guiderails are necked back down to single file, there will not be any bridging. This is an easy way to provide some inline accumulation. Figure 3.15 shows beer bottles accumulated in just such a fashion in preparation to enter a labeler.

In Figure 3.16, oil containers are shown being conveyed into a labeler (far right) from left to right. Notice that all the containers are oriented the same way so when they get to the filling machine, the spout is in the right place for the machine. In this particular application, a multiflex chain conveyor was used for the tighter turning radius.

When trying to set product into a single line, you cannot simply neck down the guiderails. This will create a situation called bridging, when two or more items are trying to go through a funnel at the same time and get jammed side by side. The pressure created by the chain running underneath holds the items in a bridge or arch similar to the stones of an arched doorway. It is far easier to get round containers into a single line than non-round containers.

When setting round containers into a single line, beaded or roller guiderails are required on one or both sides. Cold end conveyors originally developed and marketed a beaded guiderail, but now it is offered by a variety of manufacturers. The purpose

FIGURE 3.16 Tabletop chain curve.

of a beaded guiderail is that with the uneven surface, product bridging is greatly reduced or even eliminated.

A second aspect of funneling a product into a single line is the need for speed-up chains. This is the use of a series of TTC chains moving at incrementally higher speeds. Figure 3.17 shows a multi-chained conveyor with beaded guiderail on one side of the funnel. Notice that a drive is used to drive the last or fastest TTC and then a series of roller chain and sprocket slave drives are used to drive the subsequent slower chains. This type of unit is frequently referred to as a combiner (Figures 3.18, 3.19).

No hard and fast rules exist for funneling non-round containers into a single line. They require empirical testing, and it is best to deal with a company that has experience with them.

FIGURE 3.17 Multiflex system including an alpine unit feeding a labeler.

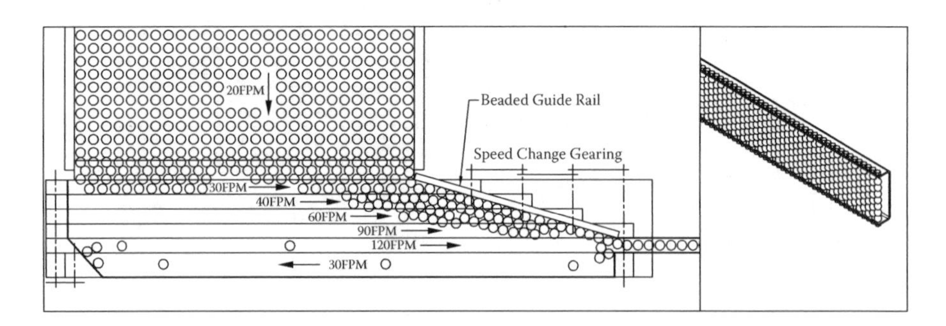

FIGURE 3.18 Combiner and the detail of a beaded Guiderail.

FIGURE 3.19 Slave drive configuration on a combiner (Guard removed).

3.4.1 Elevation Changes

From time to time, it is necessary to change elevations with TTC conveyors. Minor changes in elevation can be handled by simply inclining the conveyor for a distance. A safe rule of thumb is a rise of ⅝ in. per foot of conveyor. In other words, a 10-ft section of a conveyor can climb 6.25 in. ($10 \times .625 = 6.25$). Steeper angles can be achieved, dependent on the chain top plate material, the product material, and product stability. With steeper angles, it might be necessary to introduce a transition section so that the infeed and exit of the incline are not too abrupt. It is also important to note that a short section called a tangent is required at the infeed and exit of flat curves to ensure that the TTC runs through the curve properly. Flat curves are never inclined; only curves specifically designed as spirals should be inclined.

In the background of Figure 3.16, there is a multi-incline conveyor that is often referred to as an alpine conveyor. It is shaped like a corkscrew that has been cut in half lengthwise and stretched out. Alpines are used for elevation changes as well as inline accumulation. This is particularly useful when the product cannot be allowed to get side by side, as was the case with the oil containers shown.

When considering an incline, product orientation can play an important role in product stability. Take the oil container for instance. Referring to Figure 3.19, conveying the oil container in direction A or C, with one of the wide sides facing forward would maximize inline accumulation but would make the container prone to falling over. In this direction, there is virtually no chance of going up any kind of incline. Conveying it in direction B or D would promote stability and would certainly allow it to go up an incline better. Direction B would be best because with the neck opening leading, the container will be less likely to fall over backward.

If a product is stable at steeper angles, there is still the issue of the product beginning to slip on the chain. A point is reached where the product will sit at the bottom of the incline with chain just running underneath it. Chains with high-friction surfaces such as the one shown in Figure 3.2b are used to ensure that the product will not slip on the chain. One major concern when using this type of chain is that when product begins to back up on it, the increased friction creates a lot of line pressure and can crush or damage product. Products that need the ability to accumulate must be transferred off of the high-friction chain as soon as possible.

For more drastic changes in elevation, a side-grip elevator is in order. These units can raise or lower products at a considerable vertical distance using minimal horizontal floor space (see Figure 3.20). Some of these units are built to high enough standards that they can carry a raw egg without dropping or cracking it. There are a variety of gripper profiles. The most popular are the three- or four-fingered gripper design (see Figure 3.2c) and the D gripper design. Both are effective, but depending on the gripper material and the product being carried, one might prove to be better than the other for a particular application (Figure 3.21).

Another solution for elevation change is a spiral conveyor. Similar to the Alpine unit previously discussed, there are a few manufacturers that produce tight-radius spiral conveyors. While a couple of the variations are not specifically TTC conveyor, it is appropriate that we discuss them here. Each manufacturer has their own patented chain guiding system that makes them unique, but they serve the same purposes. These spirals can be as small and tight in diameter as the product allows, as shown in Figure 3.22a, or can be quite large to use as inline cooling or curing after a process has been performed, as shown in Figure 3.22b.

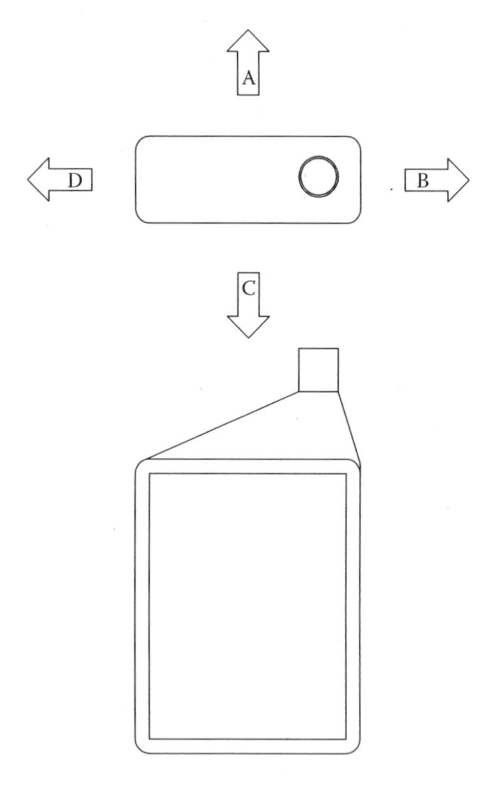

FIGURE 3.20 Oil bottle orientation.

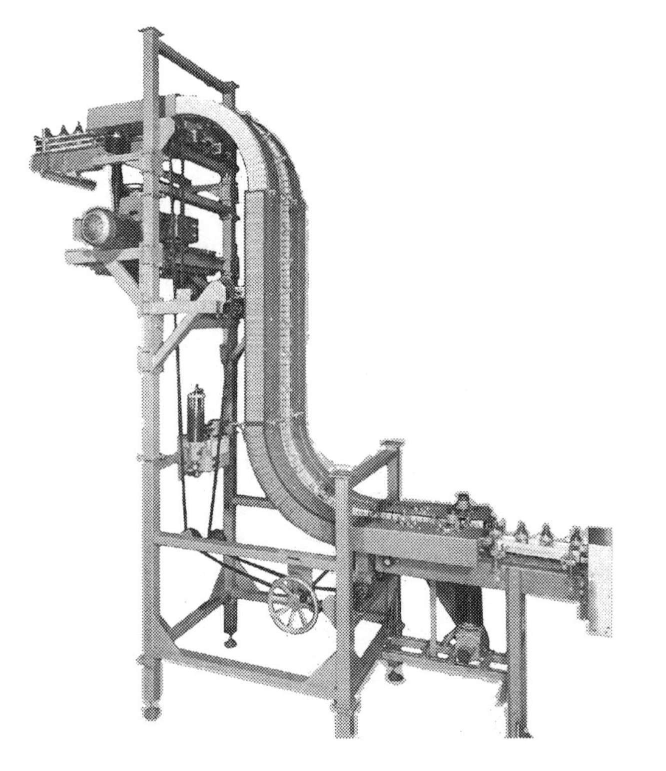

FIGURE 3.21 Side-grip elevator/lowerator.

(a) (b)

FIGURE 3.22 Spirals: (a) package handling spiral and (b) accumulating spiral.

3.4.2 Accumulation

We have frequently discussed accumulating on TTC conveyors, which is very common. Line pressure is a serious factor to be taken into account. Line pressure is the force created by the chain running under the stopped product pushing it forward. It is a function of the coefficient of friction between the product and chain. The lower the friction, the lower the line pressure. In some cases, the product cannot withstand the line pressure that can be created, so a special chain is employed that has rollers on the top plate (see Figure 3.2a). As the chain runs under the product, the rollers turn freely when the product is stopped. This type of chain, of course, is not recommended for use in any incline or decline application. The downside is that it is also not recommended for small products because it can also cause significant instability of the product causing it to tip over.

Modular plastic belting is used for wider conveyors and much larger accumulation tables. When used as a wide conveyor, these belts are treated just like the TTC conveyors. When used for accumulation tables, some special rules apply for their application. If the unit is being used for inline accumulation, there must be a reliable method for funneling the product into a single line at the exit end of the accumulation.

Now that we have discussed the various types of conveyors, let's put them to use. In Figure 3.23, product flow is from left to right. In the system shown, there are several items used. The system starts at a machine that is releasing plastic bottles onto a 4½-inch-wide TTC conveyor. That line goes through a combiner that is in front of a bidirectional accumulation table. From the combiner, they travel to a labeler that has a bypass around it. It then can go through a case packer or an elevator, and then into a bulk palletizer (a machine used to stack many layers of upright bottles on a pallet).

The major components of this system were selected for specific purposes. The accumulation table is used so that when the labeler is down for label change or minor maintenance, the machines feeding the system, such as the blow molder, do not have to shut down. The elevator is used to make the elevation change from approximately 3 ft. to 12 ft. to feed the palletizer.

For a more detailed look, first, you have to look at the product speed versus the conveyor speed. If you have 200 3-inch round cans per minute (200 × 3 in ÷ 12 in/ft.), then it equates to a product speed of 50 ft. per minute (FPM). Therefore, the takeaway conveyor must be running at at least 50 FPM. It is generally recommended that a conveyor run 20 percent faster than the product speed so that the products are not always rubbing against each other. This also allows for minor variations in production

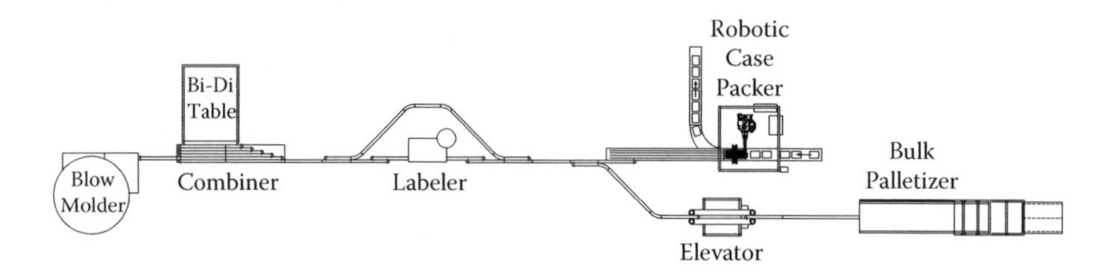

FIGURE 3.23 TTC conveyor system layout.

equipment throughput. In this case, our conveyor speed would be 50 FPM + 20 percent, or 60 FPM. This would cover the initial line. At first glance, the conveyor that is leaving the combiner should be 200 × 3 in ÷ 12 × 1.2 = 60 FPM. If there was no accumulation table there, that would be true; however, the 60 FPM only services the production rate feeding the system. After the labeler has been down and the accumulation table has product on it, the conveyor leaving the combiner needs to be running fast enough to handle incoming production as well as some extra to unload the accumulation table. For most systems, adding another 20 percent is sufficient, so 60 FPM + 20 percent = 72 FPM. This can be rounded up to an even 80 FPM. This speed can then be used throughout the remainder of the system. The other lanes on the combiner will run at 30 FPM, 40 FPM, and 50 FPM. This speed-up helps pull the containers through the combiner. With production rates forever increasing and TTC conveyors running faster, one important issue to keep in mind is that the speed differential between two adjoining lanes should not be too drastic. If necessary, additional speed-up lanes should be used. Any time product moves from one lane to another that is running faster or slower, the product will start to turn or spin. If the spinning is too fast, the product will become unstable and fall over, and any downed product can cause jams.

Accumulation tables typically run at approximately 5–10 FPM. This low speed is due to several factors:

1. The product is free-standing without any guiderails to support it and therefore more unstable as the table stops and starts.
2. The product will become unstable when it transitions onto the deadplates. The higher the speed, the greater the instability.
3. The sheer size of the table and the power required to move it smoothly.

The horsepower requirements to pull a TTC conveyor are based on the chain and wearstrip materials, the product weight and material, and the conveyor layout. This book does not cover horsepower calculations because they involve such a wide variety of factors and are best covered by the various TTC manufacturers. Chain and wearstrip materials and product data are factors, and even the overall length of the conveyor plays a role in horsepower requirements, as does the actual layout. Curves in a conveyor have a multiplying effect on the horsepower requirement. Figure 3.24 shows two layouts of a TTC conveyor. Both cover the same distance and both are the same length, but the layout in Figure 3.24b uses less horsepower because the extra length of conveyor in the lower right end of Figure 3.24a has to be pulled through two curves, whereas in Figure 3.24b, that extra length is directly at the drive and the curves have less impact on it.

In small, short conveyors, these two layouts would be virtually the same in horsepower requirements, but as the conveyor gets longer, the impact will become far greater.

3.4.3 Package Handling

With the increase in e-commerce, a lot of energy has been expended looking for faster better ways of handling packages as well as other products. This is where

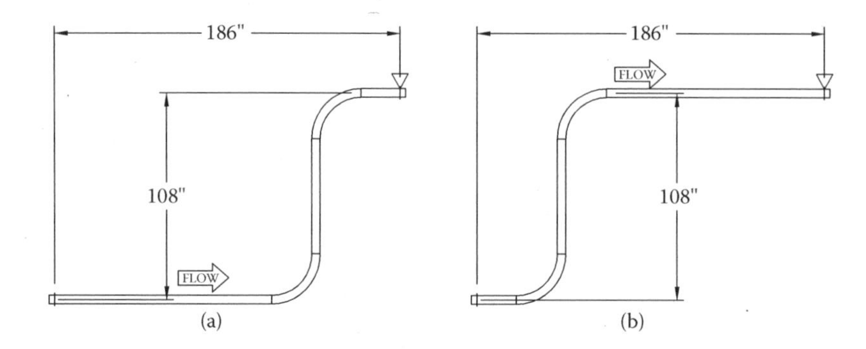

FIGURE 3.24 Conveyor example: (a) Short drive section and (b) Long drive section.

modular plastic belts have really grown in applications. With side-flexing plastic belts, it has provided the ability to carry larger quantities of product around corners, up inclines, and down declines.

Mounted roller belt (MRB) offers significantly more capabilities for package conveying and sorting. One of the greatest challenges for traditional package handling conveyor has been polybags. When you order a garment online, it is typically shipped in a plastic bag. A rectangular box with a lot of extra space would require you to pay to ship something bigger than it needs to be and/or force the garment into a properly sized box would deliver to you a very wrinkled, unattractive garment. Thus, the use of flexible mailers such as polybags. Smaller systems can't justify the expense of the types of sorters required for polybags; so, MRB conveyors make a lot of sense (Figure 3.25).

FIGURE 3.25 MRB sorter.

3.4.4 Food Handling

Modular plastic belting is very popular in the food handling industry because the available open grid design allows for easy washdown of the conveyors. In the past, open wire mesh belts were used, but with the plastic belt there is greater flexibility for the system design. The previous wire mesh belts only run in straight lines whereas the plastic belt can be straight running or side flexing (Figure 3.26).

3.4.5 Accessories

A variety of accessories are specifically designed for TTC conveyors, including manual and automatic gates, product stops, and other devices to control product flow.

As seen in the layout in Figure 3.21, prior to the labeler, there is a manual gate that allows products to bypass the labeler. A manual gate is typically used when the product path must be changed as a matter of the setup and not a function of line conditions such as a machine breakdown. Automatic gates are used so the control system can control product flow based on line conditions or to balance flow. One such automatic gate is shown in Figure 3.27.

A variety of product stops are available, each depending on the type of product to be stopped. Squeeze stops use an air cylinder to swing a short, hinged section of guiderail into the product path. These are especially effective for plastic containers. Another type of stop simply extends something into the path of the product. These work well with rigid containers, such as glass bottles or metal cans.

The last accessory discussed here is the ware controller. This device spaces out product evenly for a particular operation, such as entering an inspection machine. One such device is shown in Figure 3.27. To create the spacing, they can use a star wheel, as the device shown in Figure 3.26, or a larger, custom-designed screw (Figure 3.28).

FIGURE 3.26 Bread conveyor.

FIGURE 3.27 Air-operated gate.

FIGURE 3.28 Ware controller.

Many more accessories are available, but there is no way to sufficiently cover them all. Each has a specific use, and the conveyor manufacturer will be able to introduce you to those that best fit your specific application.

3.5 QUESTIONS

1. In Figure 3.2(a), a section of TTC that has rollers on top of it is shown. If the product to be handled will convey adequately on that chain, why wouldn't you want to use it for the entire system?
2. Which material below is not *typically* used for chain guides for curved TTC?
   a. UHMW
   b. Hardened steel
   c. Ceram-P
   d. Stainless steel
3. Given the diameter of a nail polish bottle is 1.25″ and it has straight sides, if you needed to convey them on a wide side-flexing conveyor that is about 12″ wide, what tabletop chain configuration would be best?
4. What are the product characteristics that are important when trying to select a method to raise a product from a 24″ elevation to a 120″ elevation?
5. Raised ribs on a chain or plastic belt allow product to make a smooth transfer using fingered deadplates. What would cause small product to get unstable?
   a. Deadplates mounted too low
   b. Deadplates mounted too high
   c. Chordal action of the chain
   d. All of the above
6. Which of these is not a typical top plate for plastic belts?
   a. Friction top
   b. Perforated flat top
   c. Pyramid top
   d. Raised rib
7. What is the maximum recommended elevation change achievable over a distance of 30"?
8. Why do accumulation tables run at slow speeds?
9. In Figure 3.23, a simple TTC conveyor system is shown. If the blow molder produces 100 4" bottles a minute, what would be the recommended minimum speed for the conveyor taking the bottles away from the blow molder? What would be the recommended minimum speed for the conveyor after the accumulation table?
10. What is one important issue to remember when conveying containers of product such as bottles filled with a liquid?

4 Belt Conveyors

Belt conveyors are by far the most common material handling conveyors in use today. They are usually the least-expensive powered conveyor and are capable of handling a wide array of materials. Depending on the type chosen, belt conveyors can carry everything from loose gravel and coal to disposable cameras to flimsy bags of clothing to rigid cardboard boxes full of product.

Typically, belt conveyors are used for carrying materials long distances with a single motor. Belt conveyors can range from a single unit that is several inches long up to combined units that cover hundreds of feet or even miles.

4.1 FLAT BELT CONVEYORS

Belt conveyors typically consist of a series of sections that make up an entire conveyor. Shorter units can be single self-contained conveyors. All belt conveyors consist of several sections:

- Intermediate beds
- Drive
- Take-up
- End pulleys (end brackets)
- Nose-over (RBO, optional)
- Power feeders or nose under (optional)

The purpose of each section type is discussed briefly. In a short conveyor, typically under 10-ft. length, the first four items can be encompassed within the design of a single unit.

4.1.1 Slider Bed Conveyors

Slider bed conveyors are the least-expensive type of belt conveyors and also the simplest to manufacture. As the name suggests, the belt slides over the bed of the conveyor, which is usually a formed steel section. Two common cross-sections are shown in Figures 4.1 and 4.2. The first (Figure 4.1) is most commonly used when handling packages and the second (Figure 4.2) is commonly used for airport baggage conveyors.

4.1.2 Belt-on-Roller (BOR)

Belt-on-roller conveyors are more expensive belt conveyors, but they have advantages. The BOR conveyor uses a series of rollers rather than a sliding surface to support the belt. This greatly reduces the friction between the belt and the carrying

 DOI: 10.1201/9781003376613-4

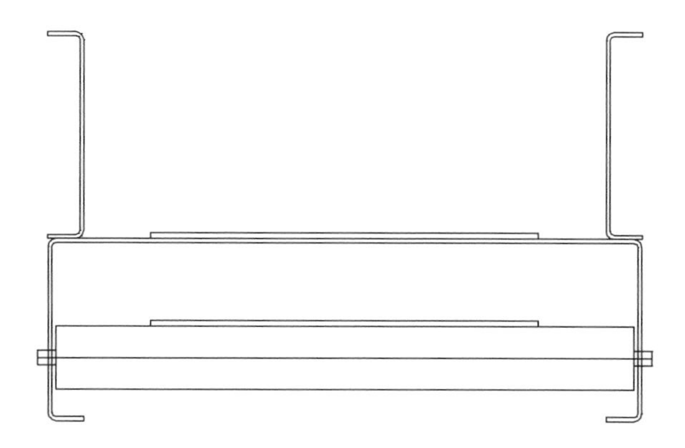

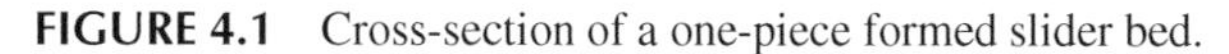

FIGURE 4.1 Cross-section of a one-piece formed slider bed.

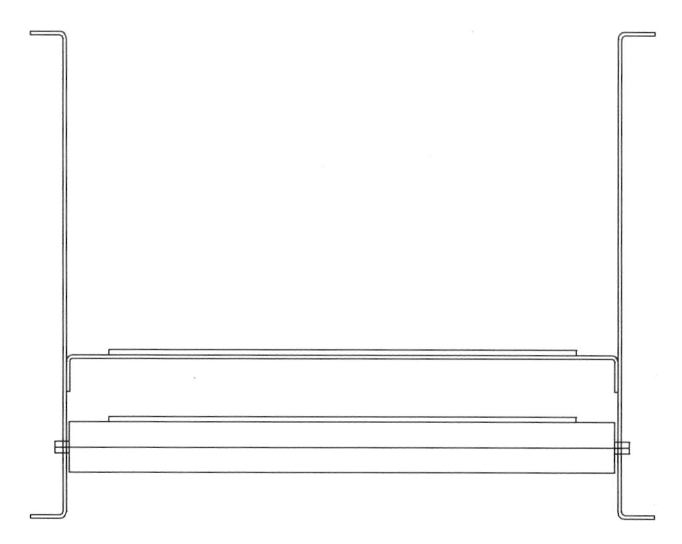

FIGURE 4.2 Cross-section of a three-piece slider bed (typically used for airport baggage).

surface. It also increases the length of conveyor or the weight of a product that a particular drive can handle. The roller spacing is based on the characteristics of the products being handled and is typically 3, 4, 6, or 9 in.

There are some drawbacks for BOR conveyors. As you can see in the cross-section in Figure 4.3, there is a gap between the edge of the belt and the side of the conveyor. Small items can fall through this gap and flexible items can get wedged under the belt. The full range of potential materials to be handled should be taken into consideration when selecting which conveyor type to use.

The BOR is shown with adjustable guard rails that are used to keep the product on the conveyor. They can also be used on the slider bed conveyor in lieu of the fixed, formed steel side pans.

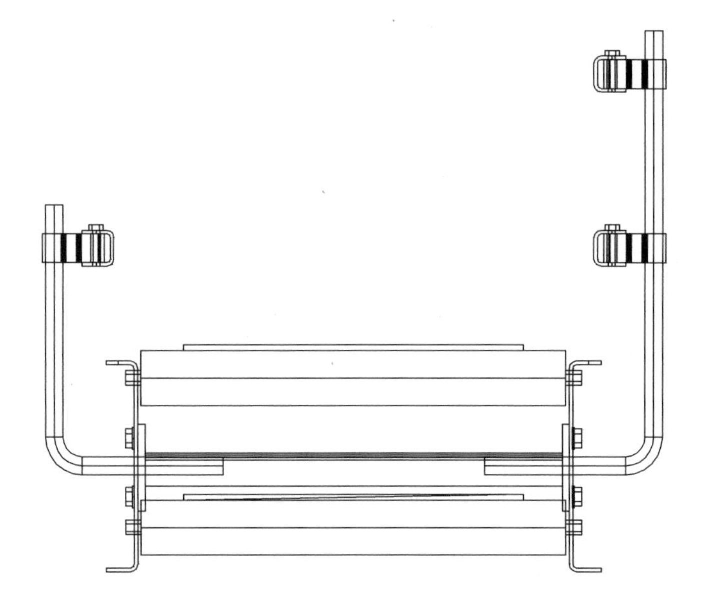

FIGURE 4.3 Cross-section of belt-on-roller conveyor.

4.1.3 Common Sections

4.1.3.1 Intermediate Beds

Intermediate sections are the simplest of the common sections. They are basically a straight bed ranging from 12-in. to 12-ft. long.

4.1.3.2 End Pulleys or End Brackets

End pulleys, as would seem obvious, are located at either ends of the conveyor to allow the belt to bend around from the top carrying surface to the return side. A snubber roller is typically included next to an end pulley. This snubber roller is used to keep the belt centered on the conveyor (see Figure 4.4).

End brackets typically use 2- to 4-inch-diameter pulleys and can also include a pop-out roller that aids in supporting small products as they transition between conveyors. The pop-out roller is also a safety device in that it allows a larger gap between the pulley and the next conveyor to minimize the possibility of injury. The roller pops out so that no one can get caught between the roller and the moving belt. If the roller is fixed in place, it can cause a wringer effect that could catch a finger or piece of clothing and pull a person into it.

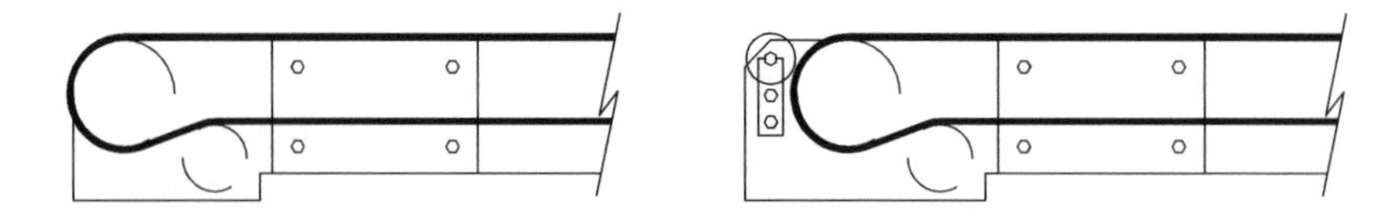

FIGURE 4.4 End pulley configurations.

Some applications for very small products require the pulley to be very small to minimize the transition gap and product instability. In these cases, a nosebar end bracket is used. Typically, these are specialty items and have a 1-in. diameter or smaller round bar or a row of bearings acting as the end pulley (see Figure 4.5).

4.1.3.3 Drive

There are two types of drives that are commonly employed: the end drive and the center or mid drive.

As the name implies, the end drive is located at the discharge end of the conveyor (see Figure 4.6). The use of an end drive eliminates the need for one of the end pulleys. The end drive is typically designed to bolt onto the end of an intermediate bed. End drives tend to be less expensive but are limited as to how long they can pull a conveyor belt. Due to the diameter of the drive pulley, a pop-out roller is frequently included to support smaller products across the transition.

Center or mid drives are located somewhere along the length of the conveyor (see Figure 4.7), ideally toward the discharge end and not necessarily exactly in the center. The center drive works best when it is within the last third of the conveyor's length. This location reduces the amount of belt under tension, thus extending the belt life. The center drive can be either an individual bed that is located between

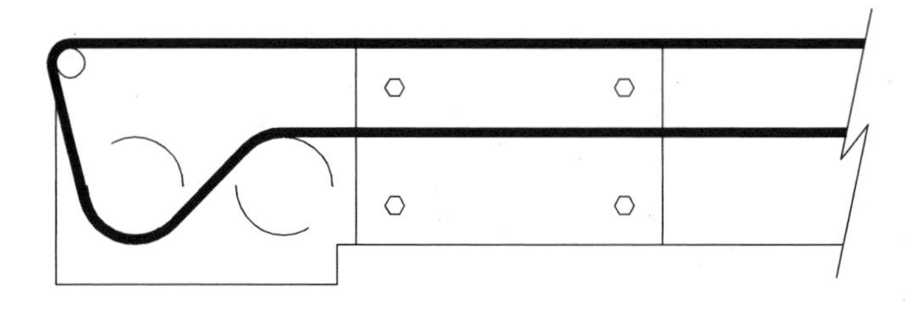

FIGURE 4.5 Nose bar configuration.

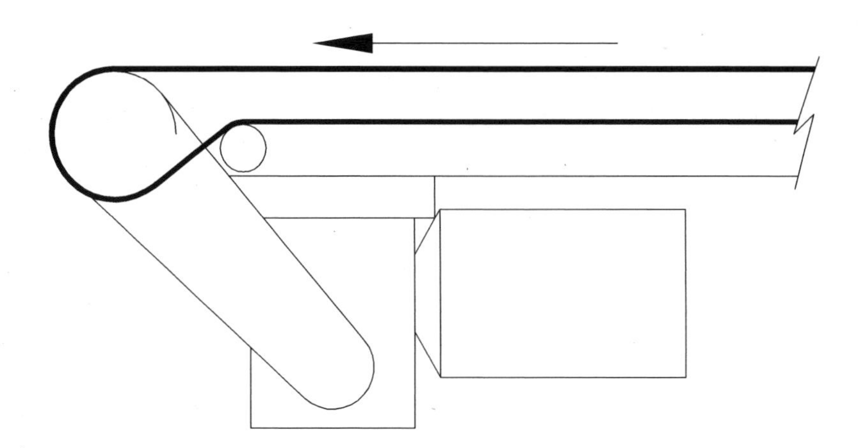

FIGURE 4.6 End drive.

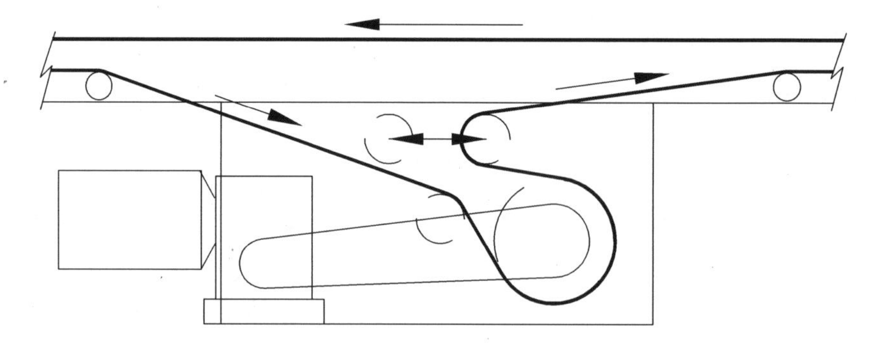

FIGURE 4.7 Center or mid drive.

intermediate beds or a stand-alone unit that is bolted onto the bottom of any intermediate bed.

Belt conveyors are driven by the friction of the belt wrapped around the drive pulley. More wrap and greater contact area provide better performance. In an end drive, the power transmitted to the belt is limited by two conditions. First, the pulley is driving the back of the belt, which is typically smooth to minimize friction on the conveyor bed surface. Second, the radius of the pulley is limited. In a center drive, the belt is wrapped around the drive pulley, sometimes with the top face of the belt in contact with the pulley. So, there is more friction and thus more power transmitted to the belt. Another advantage of the center drive is that as power requirements increase, larger pulleys can be used, thus increasing the contact area.

4.1.3.4 Take-Up

The length of take-up required for a conveyor is dictated by the stretch of the belt used and overall length of the conveyor. Most drive assemblies include take-up for belt tensioning. Where the take-up requirement is too long for the drive take-up to handle, an additional take-up unit can be added. The take-up is typically a stand-alone section that is bolted to the bottom of an intermediate bed. There are two types of take-ups: manual and automatic. The manual take-up requires someone to tension the belt periodically. The automatic take-up can employ a series of springs or air cylinders or some other means of keeping pressure on the belt. The key is to keep even pressure on the belt side to side.

4.1.3.5 Nose-Over or RBO

Nose-overs or radiused break-overs (RBOs) are employed when the belt conveyor is being used in an inclined or declined position. The nose-over is a transitional section that allows the belt to bend over from the inclined bed to an upper horizontal bed without an abrupt transition that would create a lot of friction and wear to both the belt and the conveyor beds. Figures 4.8 and 4.9 show slider bed nose-overs. Slider bed nose-overs can be either a series of flat sections with or without transition rollers between sections or a single smooth radiused section, as shown in Figure 4.10.

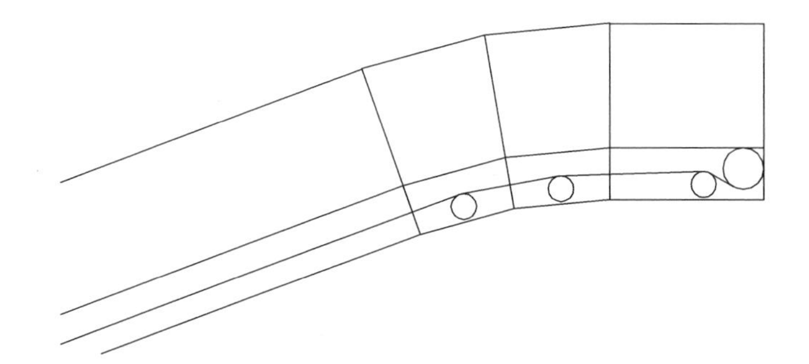

FIGURE 4.8 Slide bed nose-over without transition rollers.

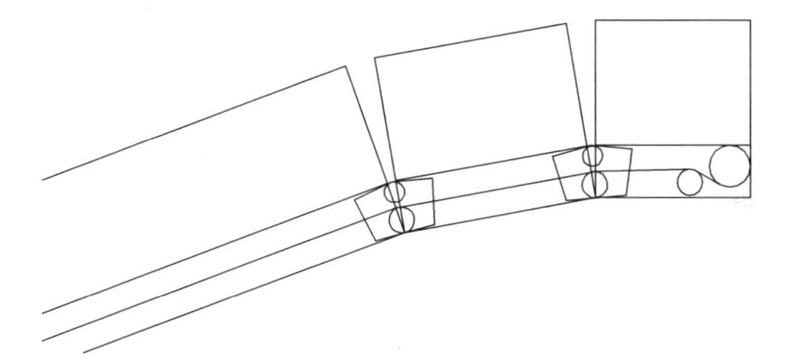

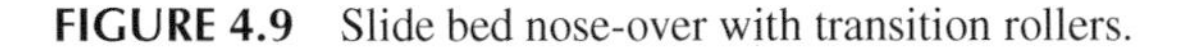

FIGURE 4.9 Slide bed nose-over with transition rollers.

4.1.3.6 Power Feeders and Nose Unders

Power feeders, like nose-overs, are used with inclined and declined conveyors. The power feeder is used when the lower end of the belt conveyor is mated to a wheel or roller-type conveyor. Power feeders, rather than having their own separate drive, are typically slave driven from the primary belt conveyor. This is done through the use of sprockets and a chain, as shown in Figure 4.11. An alternative to the slave-driven power feeder is a continuous belt that is routed around a series of pulleys, as shown in Figures 4.12 and 4.13. The power feeder or nose under provides a smooth transition from the inclined bed without the issue of the product jamming into the side of the rollers, as shown in Figure 4.14. To improve the smooth transition, a 5 percent speed increase is recommended across the transition; on an incline, the power feeder would be running 5 percent slower than the incline portion. This speed increase, of course, is only possible with the slave-driven design. Typically, a power feeder itself is not inclined more than 5 degrees.

4.1.3.7 Short Belt Conveyors

As mentioned previously, shorter belt conveyors can combine all of the required sections into a single common unit. Brake or queue belts and metering belts are two frequently used basic types of complete conveyors. Brake belts are short complete conveyors, usually around 5-ft. long. The length can vary, but there are two primary purposes for a brake belt. In a distribution system, brake belts are used at the discharge end of accumulation conveyors to hold product back from advancing

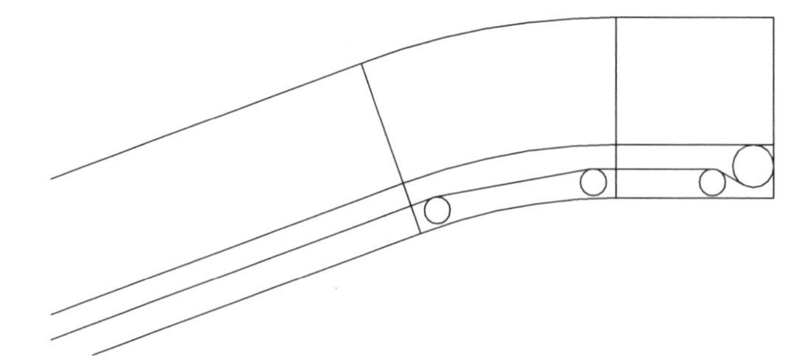

FIGURE 4.10 Radiused nose-over.

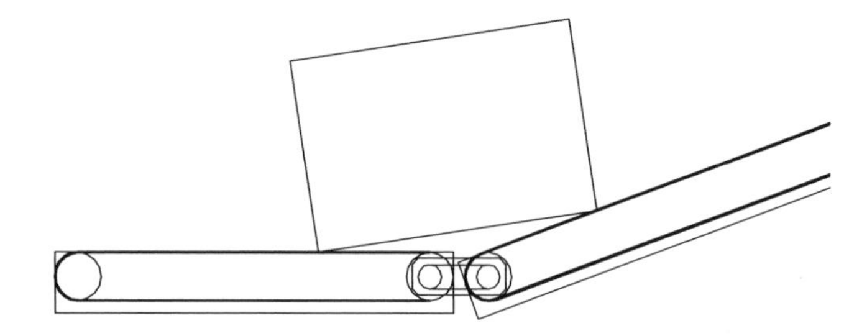

FIGURE 4.11 Slave-driven power feeder.

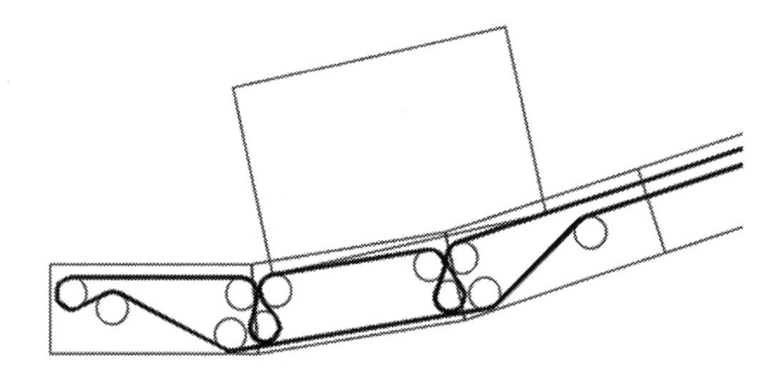

FIGURE 4.12 Continuous belt nose under.

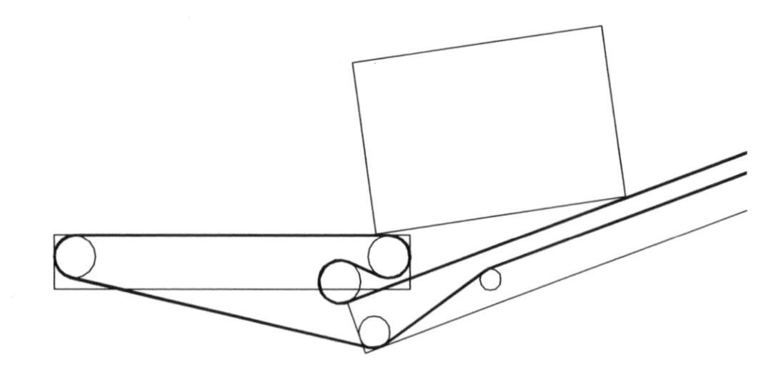

FIGURE 4.13 Double snubber belt power feeder.

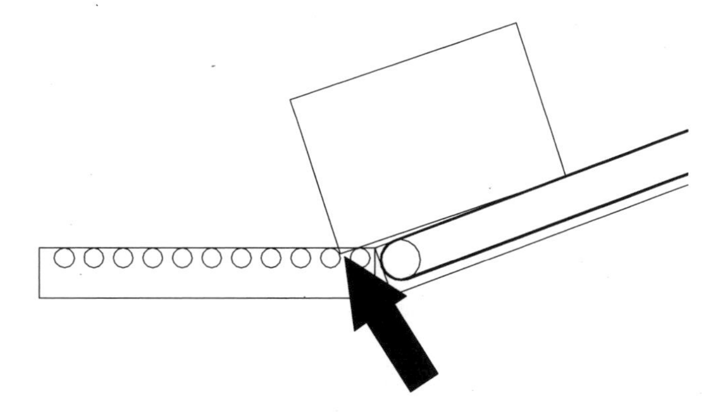

FIGURE 4.14 Product jamming into rollers during transition.

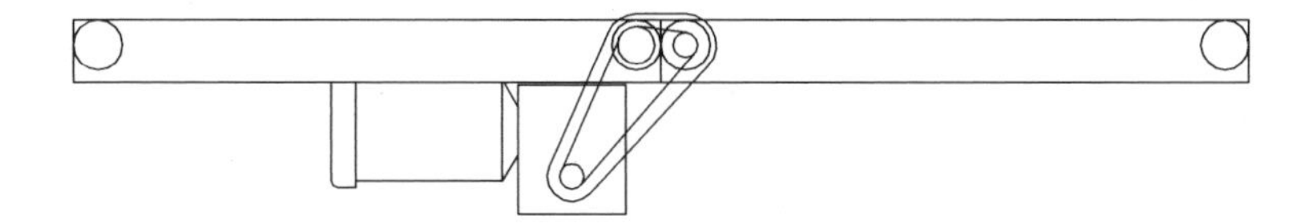

FIGURE 4.15 Metering belt.

or moving forward. Queue belts are used in airport baggage and parcel handling to individually index product through a certain process. The use of a brake belt provides more positive control of unit loads. Multiple brake belts can be placed together to create a metering belt as described next.

The metering belt (Figure 4.15) is really two short sections of complete conveyors next to each other running at different speeds. They share a common drive with the slower belt feeding the faster belt. The slower belt acts as the brake section, and the faster belt acts as the meter section. As with the brake belt already discussed, the metering belt is designed to hold unit loads back for accumulation, but then pull a gap between them as they are released over the metering portion. The speed change creates a gap between products as they transition between the two sections. This gapping is an important feature so that control devices can identify individual products as they enter a process or sortation system.

4.1.4 Application

As mentioned previously, belt conveyors are an economical means of conveying product from one point to another. When the simplest form of transportation will do the job, belt conveyors fill the need. Belt conveyors are not typically used for accumulation because, by design, products are not supposed to slip on them.

When a product must be accumulated and roller conveyors cannot be used, accumulation can be accomplished through the logic of the control system. Slug or trains of products can be indexed from a brake belt onto a longer unit that is collecting the slug. When the slug is complete, the second conveyor discharges the entire slug onto a third equal-length conveyor. The slug is broken up in basically the same manner by

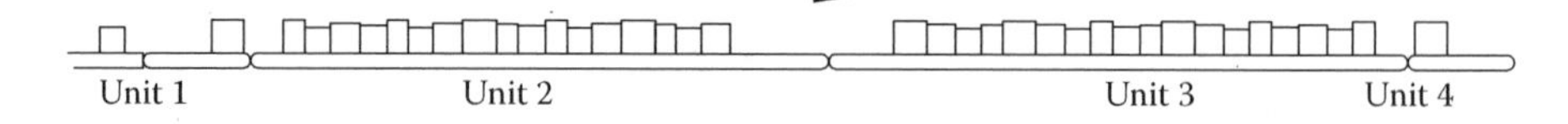

FIGURE 4.16 Slug accumulation on belt conveyors.

indexing the slug one product at a time off the end of the third conveyor (see Figure 4.16).

In selecting a drive type, there are several factors to consider:

- An end drive is less expensive than a center drive.
- Due to its design, a center drive that has the same diameter pulley as an end drive can pull a higher load. Thereby, longer conveyors can be powered by a single drive.
- With a center drive, smaller pulleys can be used on the ends of the conveyor to aid in handling smaller product.
- End drives require the least amount of clearance under the conveyor, so they have a lower minimum conveying height.
- Conveyors with a center drive can be bidirectional. Conveyors with an end drive can operate in both directions, but their capabilities are greatly reduced and belt tracking can become a serious problem.

When applying inclines and declines, it is important not to make the angle steeper than the product can safely navigate. Although it varies among manufacturers, a maximum slope of 22 degrees has been proven to be a good practice for handling most products. The relationship of height to length of a product is important in determining the maximum slope it can handle without toppling. A good rule is to make the angle of incline such that a vertical line drawn through the center of gravity of the unit load will fall within the middle third of its length as illustrated in Figure 4.17.

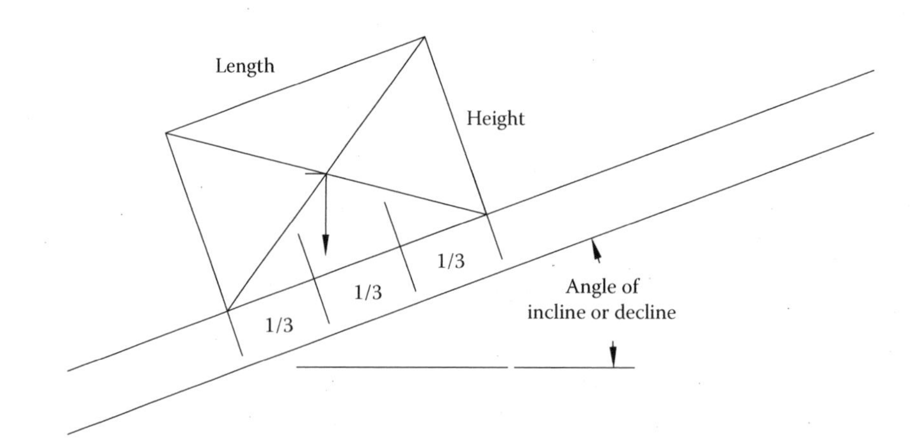

FIGURE 4.17 Relation of product center of gravity to incline angle.

Keep in mind that if the product could be top heavy, the maximum angle of incline will be less.

Figure 4.18 shows a variety of inclined conveyor applications. All of the configurations are approximately the same length and have a 20-degree incline angle, but because of their varying geometries, they have different elevation changes. Figure 4.18 illustrates basic belt conveyors, and the angle of incline or decline can be from 0 degrees (level) to approximately 22 degrees depending on the characteristics of the product being conveyed.

As mentioned previously, there are two types of flat belt conveyors: slider bed and BOR. BOR conveyors require less power to convey equal loads but are more expensive. BOR is very popular for inclined conveyors and long horizontal conveyors. Slider beds are used for declined conveyors, short conveyors such as brake belts, and when the opening along the side of the belt on BOR will cause problems.

Brake motors are also very commonly used for incline and decline conveyors as well as to keep the belt from coasting down the incline surface.

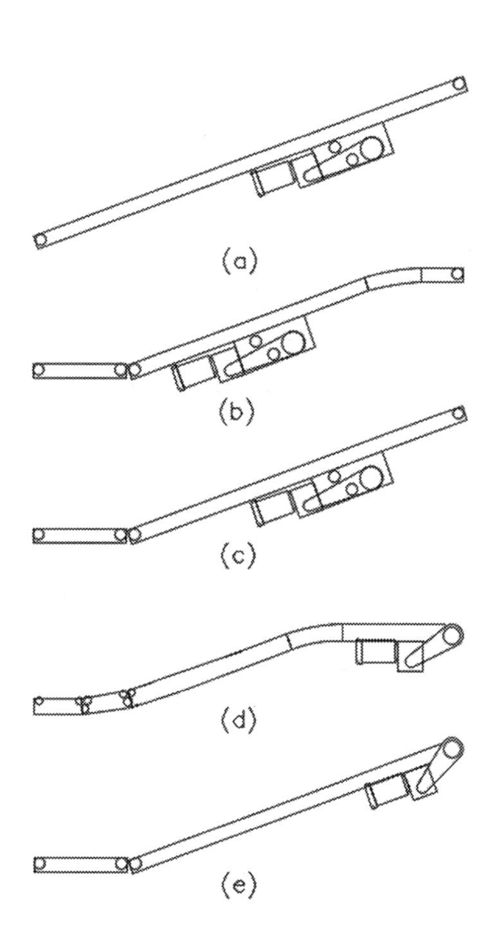

FIGURE 4.18 Various incline conveyor configurations.

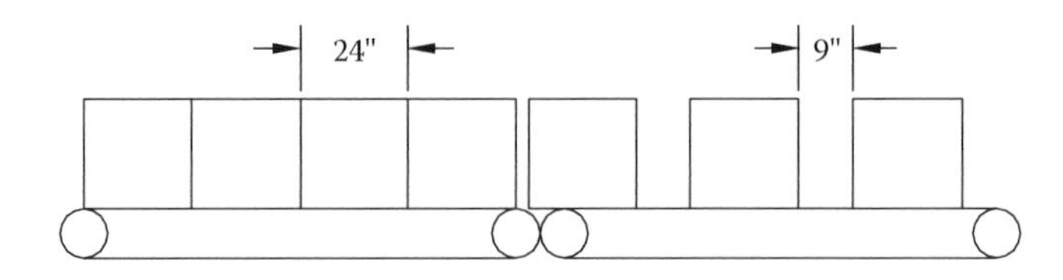

FIGURE 4.19 Product gapping using a metering belt.

Another common application for brake motors is in metering belts. As mentioned earlier, these conveyors are used to separate unit loads and put a predetermined amount of gap between them. Most sortation conveyors and transfers require unit loads to be gapped to a minimum amount to successfully divert or transfer. The metering belt does this with good accuracy. The length of the gap is dependent on the speeds of each of the belts and the length of the unit load (see Figure 4.19).

The minimum speed, measured in feet per minute (FPM), of the brake belt to meet a required rate is calculated as the average product length in feet (PL) times the required rate of cartons per minute (CPM).

To determine the speed ratio between the brake belt and the metering belt, we use the minimum product length (MPL) and the required gap (G), both measured in feet.

This ratio will ensure that there is always the minimum gap between all products. Longer products will have a longer gap, but the required rate of the average unit length will be met. One can see how critical it is to have actual unit-length data when designing a system.

4.1.5 Accessories

A variety of accessories can be used in conjunction with flat belt conveyors, and they have proven to be very helpful in certain applications. Turning wheels or posts are commonly used to aid packages in turning tight corners when transitioning from one belt conveyor to another. Pushers and swing arm diverters are used to direct product off of or onto a belt conveyor.

Guard rails or side guards are used to keep product on the conveyor as well as to "funnel" or steer the product to the middle or one side of a belt conveyor. Keep in mind that when trying to move product on a belt conveyor, it is critical to consider the friction between the belt and the product. For example, when trying to divert a product across a belt conveyor using a fixed deflector, the belt should allow the product to slide enough so that it does not affect the tracking of the belt.

4.2 BELT CURVES

When a change in direction to the left or to the right is required, a belt curve conveyor can be used. Like straight belt conveyors, they are used to convey products of varying sizes, shapes, and weights. Because the conveying medium is a flat belt, these conveyors are capable of moving everything from well-formed boxes, cartons, and totes to very flimsy bags and bundles. Belt curves, however, are not recommended for bulk material applications.

Belt curve conveyors are typically flat, operate at a single elevation, and can be supplied in almost any arc, depending on system layout requirements. Typically, all flat turns have a single drive for up to 180 degrees. When an elevation change is required, whether up or down, a helix or spiral belt curve is used. If a larger turn is required, it is broken into multiple units. For some spiral belt curves, depending on outside conveying radius (discussed next) and product loading, arcs as high as 450 degrees have been achieved.

4.2.1 Design

Belt curve conveyors are typically shipped as complete units. For logistical reasons, curves can be made in multiple sections. The rarity of this does not require a distinction to be made between drive and intermediate sections. Therefore, we discuss belt curves as complete units.

Belt curve conveyors are categorized into several size groups based on the outside radius of the conveying surface:

A	Up to 1,200 mm (47 in.)
B	1,201 mm to 1,500 mm (59 in.)
C	1,501 mm to 2,200 mm (87 in.)
D or S	Over 2,200 mm (87 in.)

Some manufacturers offer only a limited number of outside radii coupled with choices of belt widths. For instance, they might offer a 1,500 mm outside radius with 600 mm, 800 mm, or 1,000 mm belt widths. Depending on the manufacturer, virtually any radius and belt width combination is available. There are some geometric limitations such as pulley diameter that are discussed later.

4.2.1.1 Frame Construction

Whereas straight belt conveyors are available as either slider bed or roller bed, belt curves almost always utilize a slider bed concept. The sliding surface is typically a low-friction surface of unpainted steel, galvanized steel, or phenolic-impregnated wood composite. In instances where reduced friction in heavy load applications is desirable, rollers or wheels are used in conjunction with the slider bed.

There are several prominent frame design types. The first and most common is a welded steel frame. The majority of manufacturers use this type because it is the easiest to manufacture with the inside and outside frame members welded to the top plate to create three sides of a box frame (see Figure 4.20). This type of frame also lends itself to stainless steel washdown applications due to the sealed box structure.

The next frame design is the open frame (see Figure 4.21). The open frame can be either welded steel or extruded aluminum. The open design allows for easier visual inspection and access to some service parts.

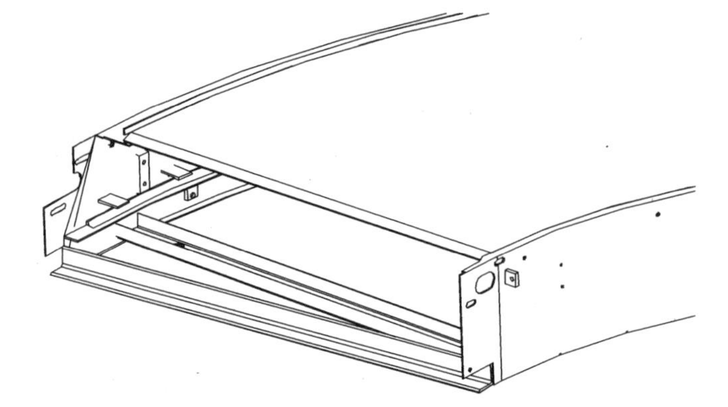

FIGURE 4.20 Welded steel frame of a belt turn.

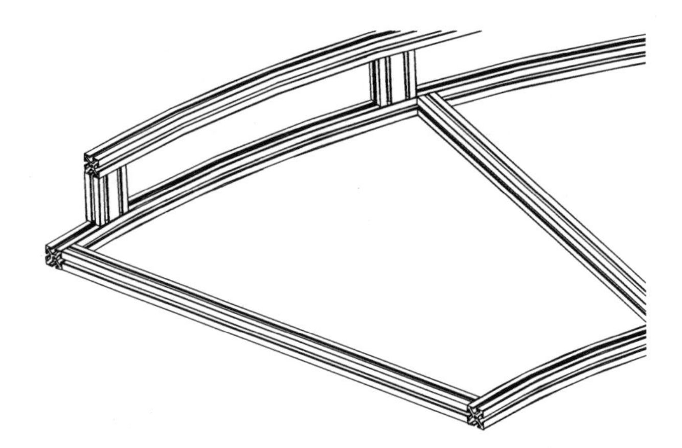

FIGURE 4.21 Open extruded aluminum frame of a belt turn.

4.2.1.2 Drives

The primary drive types available on belt curves are friction drives, pinch drives, and ancillary drives. This refers specifically to how the belt is driven rather than the location of the drive as with straight belt conveyors. Each of these drive types dictates where the drive is located on the curve.

The friction drive directly drives the belt the same way a straight belt conveyor drives the belt. Conveyors that utilize a friction drive are engineered with pulleys tapered to exacting tolerances to ensure good belt traction. Friction drives can operate reliably at high speeds, such as 600 FPM or higher. Because the low-friction bottom of the belt is in contact with the pulley, there is a limit to the load a friction drive can carry. The belt of a friction drive has a guiding element added to the outside periphery of the belt. This guiding element can be a series of bearings or a guide bead. The guide bead is typically sewn or riveted on and provides much quieter operation than the bearings. Although not recommended, friction drives can be used to push the belt rather than the typical arrangement of pulling the belt through the curve.

FIGURE 4.22 Typical ancillary or chain drive.

The pinch drive squeezes the belt between two rotating wheels. This drive allows for looser tolerances on the tapered pulleys because they are simply used to track the belt. Pinch drives typically have the lowest driving capacity of the three types. Like the friction drive, the belt typically has a guiding element added to the outside periphery. An advantage of the pinch drive is that it can be reversed without modification to the conveyor. Pinch drives are typically located at the center of the outside arc so that when they are reversed they are never pushing the belt through the curve.

The ancillary or chain drive pulls a chain or other power transmission device that is attached to the belt and pulls the belt along with it (see Figure 4.22). Ancillary drives are more forgiving of the tolerances of the tapered pulley. The ancillary drive can pull heavier loads than either the friction or pinch drive, and in some cases, can operate reliably at high speeds such as 600 FPM or higher. Ancillary drives that utilize side-bow chains as the driving force require that the chain, drive sprockets, and chain guide be lubricated regularly to maintain optimum operation and minimize wear. An ancillary drive can also be used to push a belt, depending on speed and loading requirements. Depending on the chain guide design and belt tension, side-bow chain can tend to jump the sprocket teeth and accelerate their wear when pushing the belt. Depending on the frequency of periodic maintenance, chain drives will become noisier over time.

4.2.1.3 Belts

On straight-running belt conveyors, it is preferable that the belt be stiff across the width of the conveyor to keep it flat and keep product from getting under the edges of the belt. These belts utilize a monofilament weft and multifilament warp (see Figure 4.23). Curved belts, however, require what is referred to as a balanced weave. When a curved belt is cut from a roll of belting, the resulting piece has the warp and weft pointing in different orientations throughout the arc of the belt (see Figure 4.24). Because the monofilament weft stretches differently than the multifilament warp, a balanced weave typically employs a multifilament construction for both the warp and the weft.

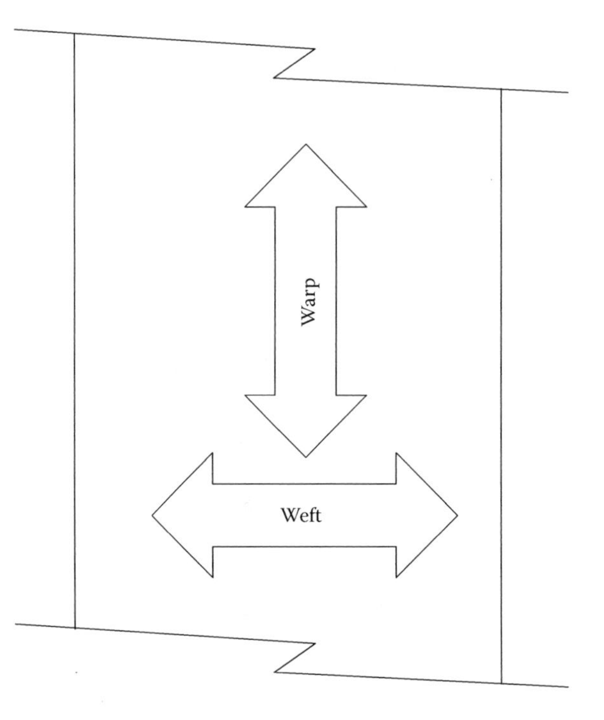

FIGURE 4.23 Direction of warp and weft in belt material.

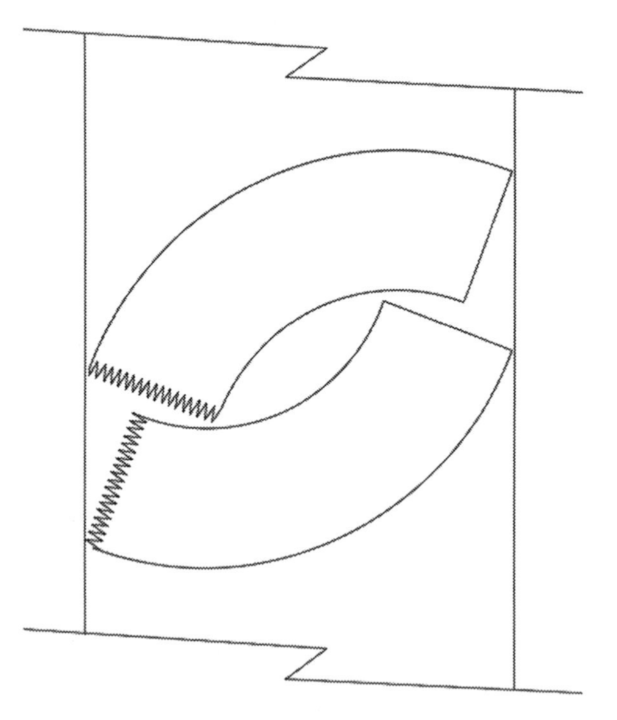

FIGURE 4.24 How curved belts are cut from straight belting.

The belt design is closely tied to the frame design and drive type. Due to the basic principles behind a belt curve, the belt has a natural tendency to pull toward the center. An integral part of the belt design is how the belt is held in place. The primary belt design options include three primary choices: beading, chain, or bearings.

Beading utilizes a urethane bead that is sewn onto the outer perimeter of the belt (see Figure 4.25). This is typically used with friction or pinch drive designs. The beading is held to the outside through a series of bearing holders. With this design, there is a close relationship between the beading and the belt; it is important that the belt be of a design that has a balanced weave and tracks well through the curve.

As shown in Figure 4.26, in the ancillary drive design, the chain is attached to the belt through a series of rivets. The chain travels through a guide track that holds it out at the proper radius.

The third popular variation is a series of bearings attached to the belt around the outer periphery (see Figure 4.27). The bearings are generally bolted on either side of the chain through a rivet or grommet. Similar to the beading, this is typically used with the friction or pinch drive designs.

The beading design, although offering the quietest operation, does have limitations when handling particularly heavy loads. The chain design, along with the bearing version, offers the highest load-handling capacity but might not be as quiet as the beading design at higher speeds. Because the ancillary and bearing designs actually hold the belt in place and have little dependence on belt tracking, it is important that the chain or bearing guidance system be robust.

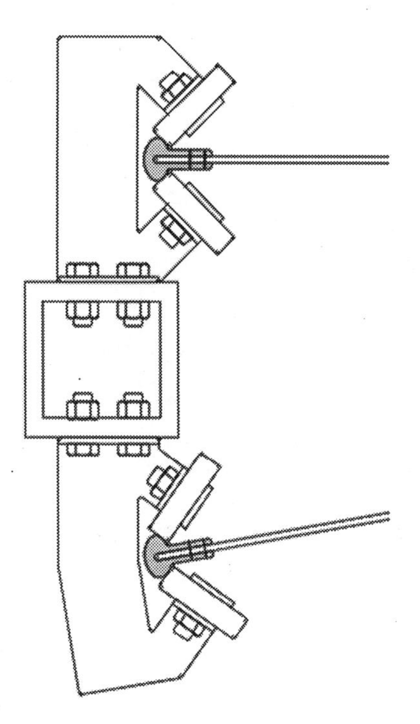

FIGURE 4.25 Bead-type belt retention.

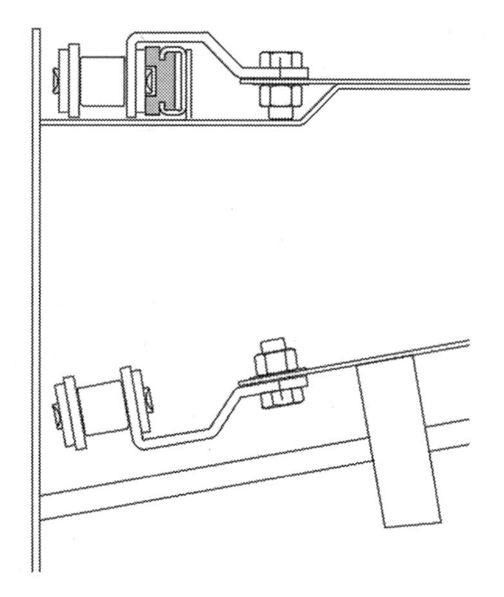

FIGURE 4.26 Chain-type belt retention.

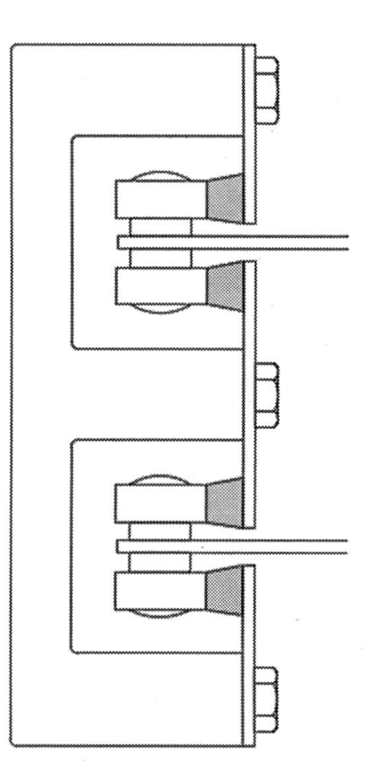

FIGURE 4.27 Bearing-type belt retention.

Belts either can be endless with a vulcanized splice or can utilize a clipper lacing. The clipper lacing simplifies belt replacement, and the endless belt ensures a continuous surface with no catch points.

4.2.1.4 Pulleys

Pulleys on belt turns are typically tapered to match the radius of the curve. The taper is designed to ensure proper belt tracking. The pulley can be made in a variety of ways and from a variety of materials. Pulley constructions include the following variations:

- Roll-formed tapered steel
- Cold-formed steel
- Cast urethane
- Machined urethane
- Turned wood

In the case of the steel pulleys, a rubber or urethane lagging may be added for drive pulleys for additional traction on friction-driven turns (see Figure 4.28).

Chain drives require that a sprocket be added to the same shaft as the pulley to drive the attached chain or other power transmission device (see Figure 4.29).

4.2.2 Application

4.2.2.1 Determine Curve Radius and Belt Width

The width and radius are selected based on the longest and widest unit load to be handled. The following is a formula to determine the spacing required between the side guards (see Figures 4.30, 4.31):

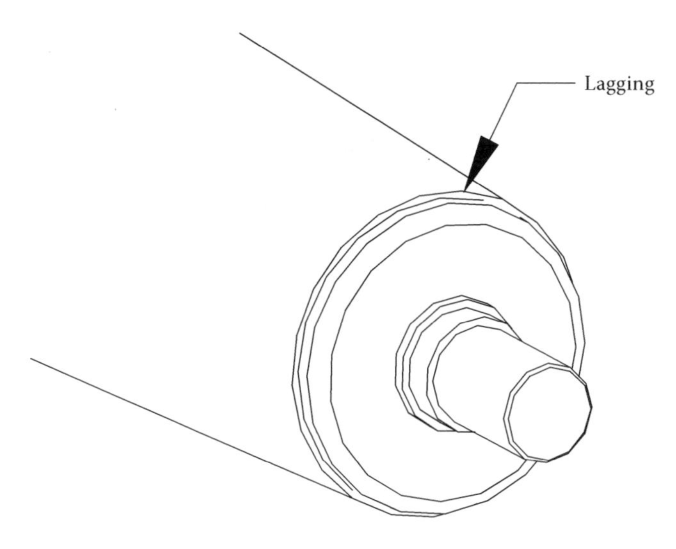

FIGURE 4.28 Urethane lagging on a steel pulley.

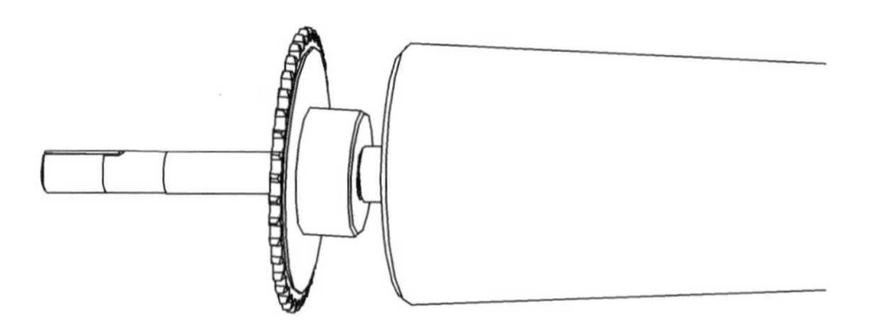

FIGURE 4.29 Ancillary or chain drive pulley.

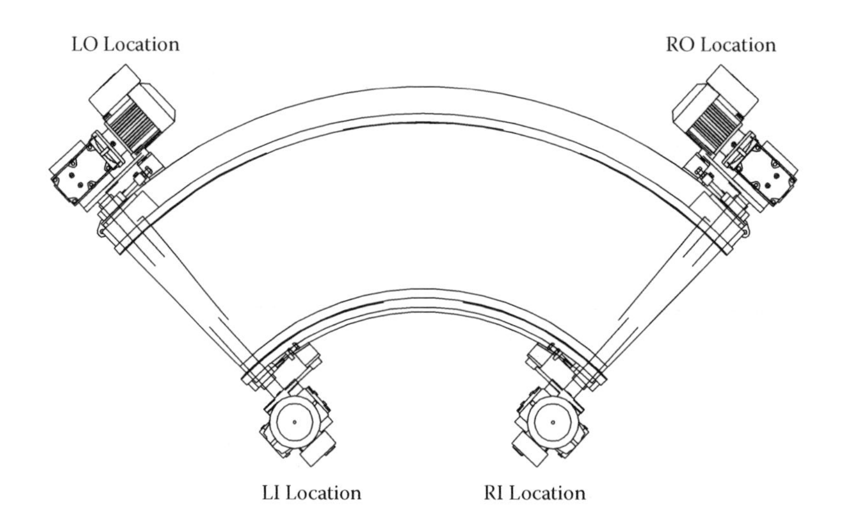

FIGURE 4.30 Belt turn drive location options.

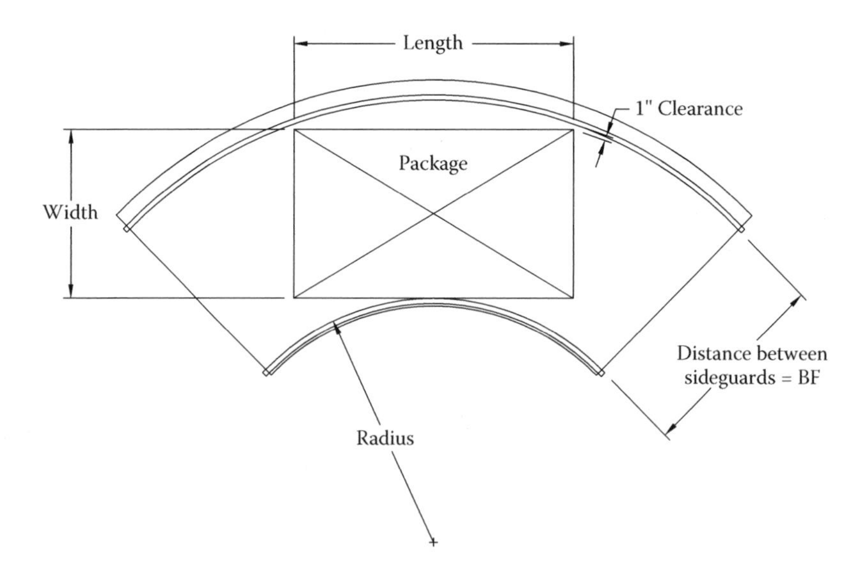

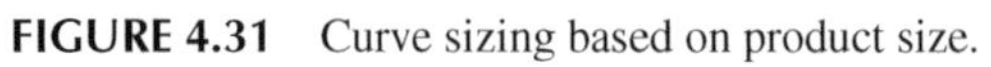

FIGURE 4.31 Curve sizing based on product size.

$$BF = \sqrt{(Radius + PackageWidth)^2 + \frac{PackageLength^2}{2}} - Radius + 1$$

4.2.2.2 Determine Incline for Helix Curves

Just like straight belt conveyors, there are limitations to how steep a helix belt curve can be. A good rule of thumb is that the angle of incline at the inside radius should not exceed 22 degrees. This number may increase or decrease depending on the type of belt used and the physical characteristics of the product being handled.

The angle of incline (α) is always steepest at the inside radius (IR). Use the following formula to determine the angle of incline based on the height difference (HD) and the sweep angle (A) of the curve (see Figure 4.32):

$$\alpha = \arctan\frac{HD \times 180}{IR \times \Pi \times A}$$

The geometry of a helix belt turn dictates that at a given centerline radius and height difference, as the belt width increases, the angle of incline on the IR also increases. This plays an important role when dealing with a variety of product sizes. With the curve wide enough to handle the largest product, the IR should not be too steep to safely handle the smallest product.

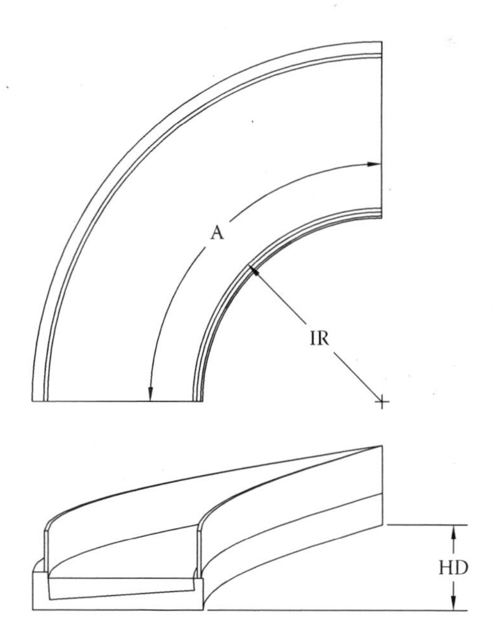

FIGURE 4.32 Geometry of a helix or spiral belt turn.

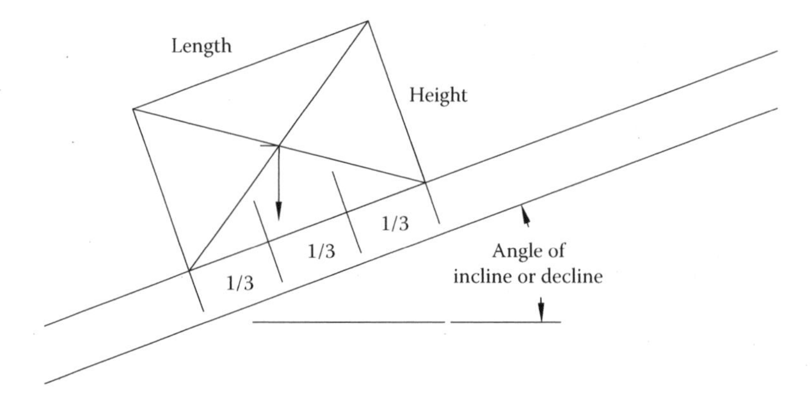

FIGURE 4.33 Relation of product center of gravity to incline angle.

Figure 4.33 is an illustration that shows how to determine the maximum angle of incline that a unit load can handle without toppling. This is identical to Figure 4.17. The relationship of height to length of a unit load is important in determining the maximum slope it can traverse consistently. The safe rule is to make the slope such that a vertical line drawn through the center of gravity of the unit load will fall within the middle third of its length.

4.2.3 Accessories

Typically belt curves are used for transportation; therefore, the accessories available are primarily concerned with improving unit load handling. Transfer brushes and pop-out rollers are two accessories that help support unit load as it transitions from one conveyor to the next.

Side guards are very common and are most frequently made from formed and rolled sheet metal. Channel type and extruded ultra-high-molecular-weight polyethylene (UHMW) have also been used to guide unit loads on belt curves.

In airport baggage and parcel handling applications, there are a variety of pushers and diverters that are available. The pushers and diverters are all motor-driven rather than pneumatic- or hydraulic-powered. This is because the varying and sometimes extreme environmental conditions around an airport are not conducive to the use of pneumatics. With hydraulic systems, if a hydraulic line leaked or broke, the fluid could conceivably get on the baggage being conveyed.

There are two primary types of pushers: what is referred to as the Boeing Airport Equipment (BAE) style and the rotary pusher. Pushers have been used for many years in the airport baggage industry, but there is a relatively new and far more popular method being used to divert bags. This is through the use of one or two vertically mounted belts that swing out into the path of the baggage, forcing it off the side of the conveyor, either onto another conveyor or down a chute (see Figure 4.34). These new diverters activate faster and are gentler to the baggage.

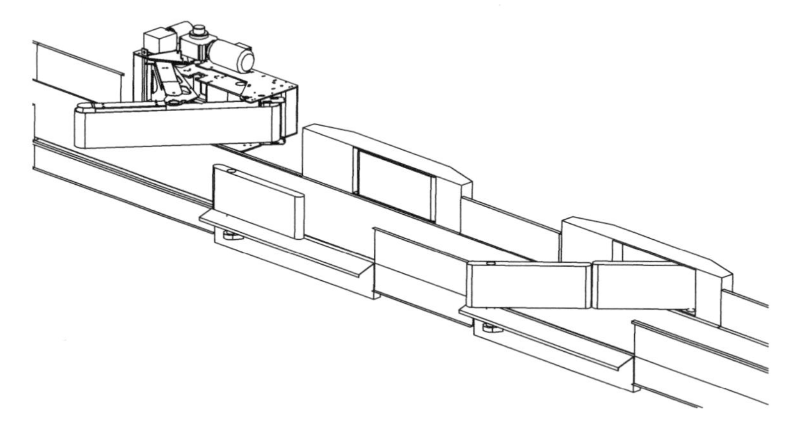

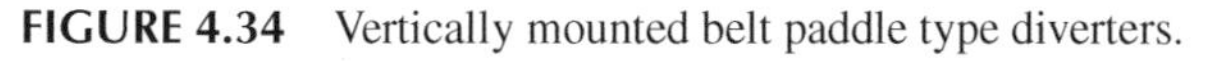

FIGURE 4.34 Vertically mounted belt paddle type diverters.

4.3 TRAILER LOADER/UNLOADERS

A group of specialty conveyors called loader/unloaders are used to load and unload trucks and trailers. There are two primary types: roll-out and extendable loader/unloaders.

4.3.1 Roll-Out Trailer Loader

With a roll-out trailer loader, the entire belt conveyor is mounted on wheels and rolls into the truck or trailer. The wheels can be motor driven to assist in moving the unit forward and back. Because the unit is on wheels, if a trailer is not level, the unit follows the slope of the trailer. This unit is used almost exclusively for trailer loading rather than unloading. Units can be equipped with a gravity discharge section on the end of the loader. As the trailer or truck is filled, the loader is backed out (see Figure 4.35). The drawback to this type of loader is that when the unit is parked in its unused position, it takes up the full conveyor length in floor space.

4.3.2 Cantilevered Extendable Belt Loader/Unloader

Cantilevered extendables are securely mounted on the dock and telescope out into trucks. Extendables are made up of multiple sections or booms that extend out from the base (see Figure 4.36). The drive for the extension feature and the belt are housed in the base of the unit.

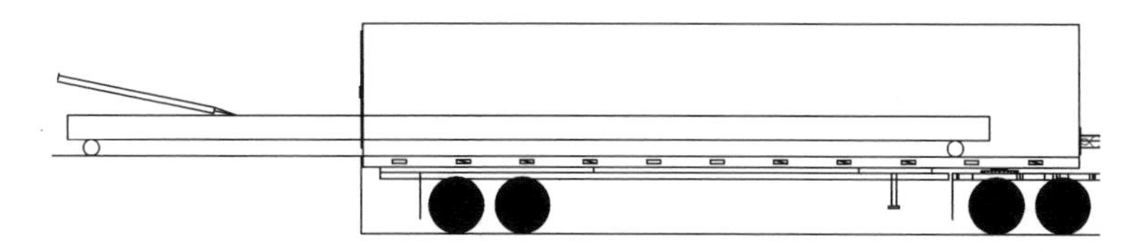

FIGURE 4.35 Roll-out trailer loader.

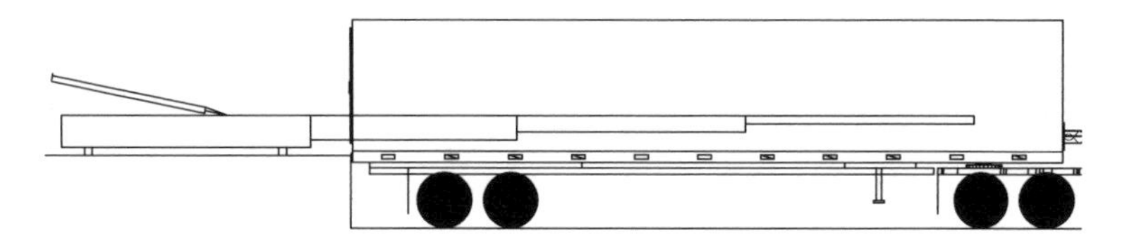

FIGURE 4.36 Extendable belt conveyor.

There are several ergonomic features available as well. By adding hydraulic cylinders to the rear of the base, the entire unit can be tilted up or down to suit the operator. Another ergonomic feature is a separate short belt conveyor mounted to the end of the extendable called an articulating boom extension (ABE) or a droop snoot. The ABE has separate controls to allow the operator to move it anywhere that is more convenient and comfortable for him or her to work.

WARNING: It is important to note that the operator end of any powered loader should have an emergency shutoff so that an operator cannot be trapped between the conveyor and the side of the truck or the product already in the truck.

4.4 SPECIALTY FLAT BELT CONVEYORS

Like all conveyor groups, there are a variety of specialty belt conveyors. Typically, we do not discuss them at great length. Due to the sheer number of them, we do not attempt to list them all.

Two specialty belt conveyors we do discuss are angled merges and strip merges. They are used extensively in airport baggage, as well as parcel and package handling. The idea is to introduce product to a transfer point at an angle to better facilitate the transition. The angles at which they are designed are typically 45 degrees, 30 degrees, and 20 degrees, but that does not mean other angles cannot be achieved. The standard angles are adhered to because there is a fairly high level of engineering involved in making a new variation, and it is typically not justifiable.

The angled merge uses a single flat belt that is wrapped at an angle around a nosebar (see Figure 4.37). The belt follows a path similar to a conveyor that has been turned inside out at one end and folded in the middle. The nosebars can be as small as 1 in. in diameter or as large as 4 in. Obviously, the smaller the nosebar, the closer the angled face can get to another conveyor. The drawback is that as the nosebar shrinks in size, the friction and wear on the belt increases, so there is a trade-off. Regardless of the nosebar type, it is important to keep in mind that it is a serviceable item that must be replaced and the conveyor chosen must make allowances for that change. Angled merges are typically available in 30-degree and 45-degree configurations. These are most often used with airport baggage applications or with larger products.

The strip merge serves a similar purpose but uses a series of narrow belts rather than one flat belt (see Figure 4.38). The strip merge is more expensive but has no nosebar to deal with. The key to strip merges is that each belt must be individually guided and have its own means of tensioning. Many designs are available, and it is important to use one that fits the application and is serviceable. It is very impractical

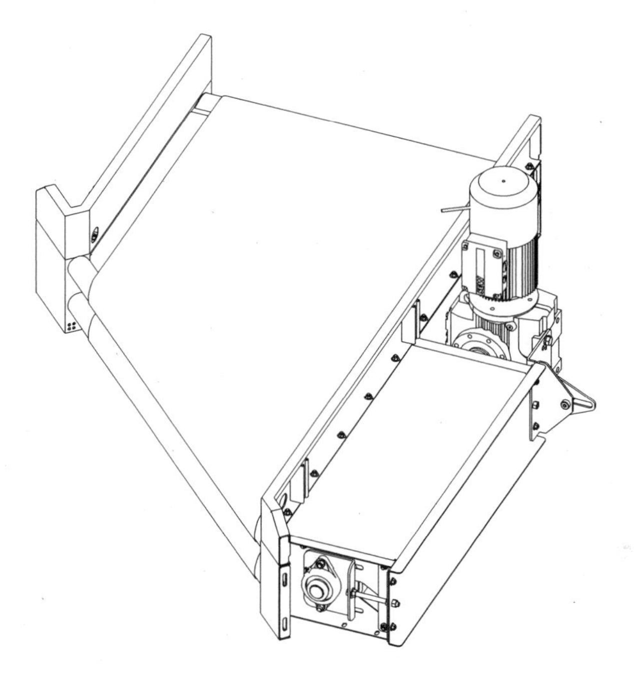

FIGURE 4.37 Forty-five-degree merge conveyor.

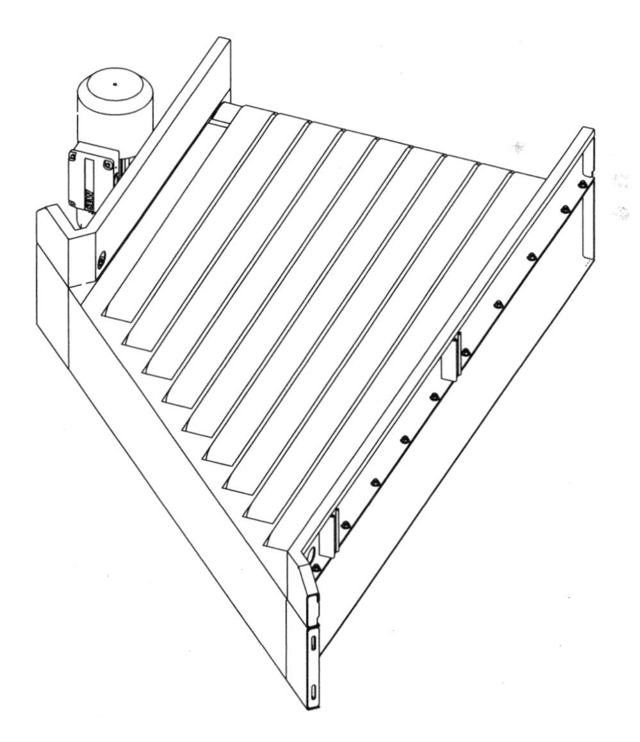

FIGURE 4.38 Strip merge conveyor.

to have to disassemble the entire conveyor to replace one belt should it get damaged. Depending on the products to be handled, the deck that is visible between the belts should be designed to prevent thin product such as envelopes from getting under the belts. Because of the high cost of strip merges, they are typically only provided when an angled merge will not work effectively. Strip merges are typically available in 20-degree and 30-degree configurations. Strip merges are most frequently used with relatively small or lightweight products, such as postal applications, and as an induction belt for tilt tray and cross-belt sorters.

4.5 CLEATED BELT CONVEYORS

This group of conveyors uses cleated belts to aid in the conveying of product. Frequently, they are used as under-press conveyors and are designed to sit under the press of injection molding machines to catch plastic parts as they are ejected. A normal flat belt conveyor with enclosed sides can be used for this application, but in many applications, especially where small parts are involved, a special type of belt is used. With this special belt, certain conveyor features are required.

Cleated belt conveyors typically feature cleated belts and integral sideboards rather than guard rails. Cleats are small barriers that are attached to the top of the belt so that as the belt goes up an incline, the product rests against the cleat and is carried up. Cleats greatly increase the maximum incline angle at which a conveyor can effectively operate.

Cleats can be attached to the belt in a variety of ways. The two most popular are bolted on and vulcanized, as shown in Figure 4.39. Both have advantages and disadvantages.

Bolt-on cleats can travel over smaller pulleys because they do not alter the flexibility of the host belt. Bolt-on cleats are also easily replaced when damaged. The primary drawback is that small parts can get caught between the belt and the cleat.

Vulcanized cleats provide a seamless joint between the belt and the cleat so nothing can get underneath. Because the cleat alters the flexibility of the belt, however, it cannot wrap around smaller pulleys. Additionally, if a cleat gets damaged, the entire belt must be replaced.

Cleated belts are custom fabricated to order, so they can be ordered in different heights, different materials, and with different spacing between cleats. Some

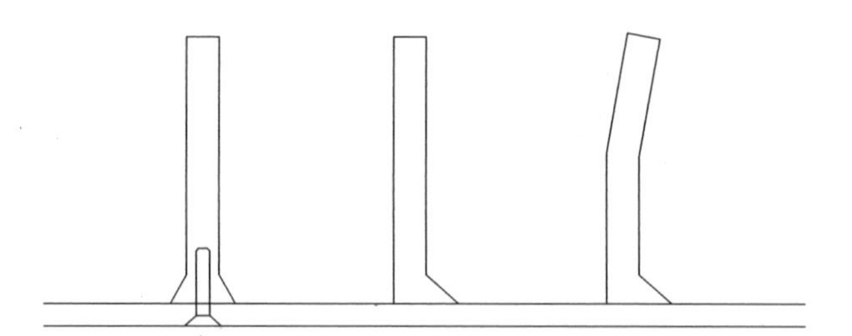

FIGURE 4.39 Sample belt cleat profiles.

manufacturers offer various cleat profiles based on the application, the weight and size of the product being handled, and the angle of incline.

Funneled sideboards are used to catch the parts and direct them toward the center of the belt (see Figure 4.40). In lieu of sideboards, corrugated sidewalls can be used on the belt. The corrugated sidewalls shown in Figure 4.41 are wavy longitudinal edge cleats designed to expand when flexing around pulleys, ideal for containing very fine free-flowing parts or materials.

To this point, we have been discussing traditional cleated belts. Now we want to introduce another type of belting. Plastic belting, as discussed in Chapter 3, is made up of a series of molded plastic links that are assembled in a brick pattern. Most manufacturers of this type of belting also offer belt sections with cleats and sideboards. The typical pulleys are replaced by a series of sprockets, but the rest of the conveyor remains virtually unchanged. This type of belting can be used for the same applications as the traditional belt, but it can cost more. Plastic belting can be designed to allow air to pass through it. We have used it as a cooling conveyor for plastic parts. The parts drop out of an injection molding machine and onto the cleated

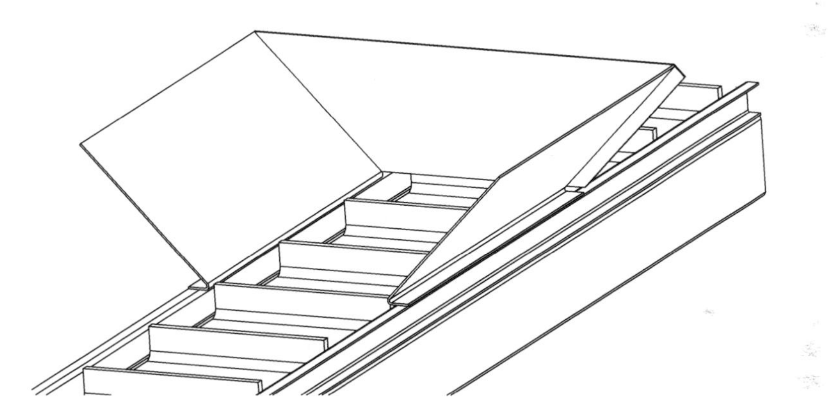

FIGURE 4.40 Cleated belt with funneled chute and fixed sidewalls.

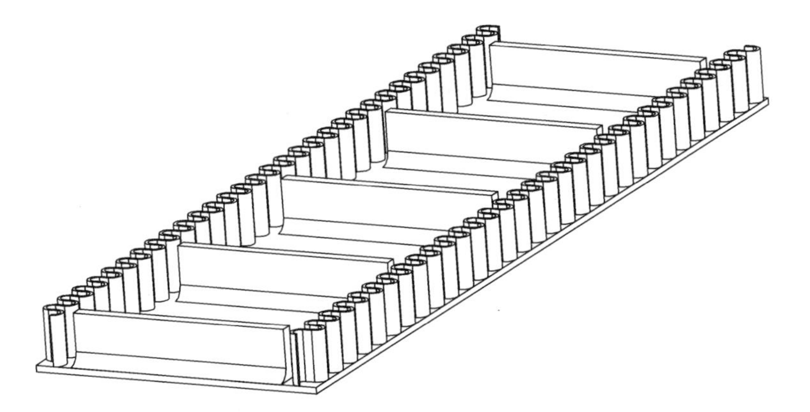

FIGURE 4.41 Cleated belt with corrugated sidewalls.

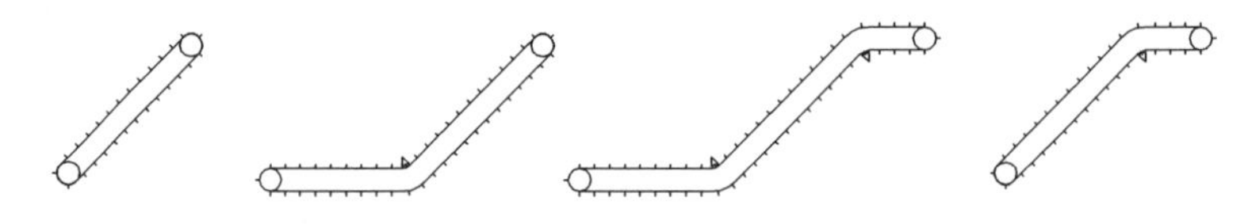

FIGURE 4.42 Various under-press conveyor configurations.

belt. While they are being conveyed up and into a large storage box, chilled air is being forced up through the open mesh of the belt to cool the parts so that they are fully cured by the time they are dropped into the container. This way the parts on the bottom do not get crushed out of shape due to the weight of the others on top of them.

The components of a cleated belt conveyor are the same as other belt conveyors, but there are three primary differences. One is that there are no return rollers on the underside of the conveyor because the cleats would get hung up on them. Instead, a formed steel pan or strips of extruded UHMW are used to support the belt on the return side of the conveyor.

The second difference is that, again due to the cleats, center drives are not used because the face of the belt cannot be wrapped around a pulley. All cleated belt conveyors use end drives.

The final difference is the use of a hold-down device at the point of transition between horizontal and incline. The hold-down device only holds the outer edge of the belt, rather than the full width, so this greatly limits the maximum pull that the conveyor can handle before the belt jumps out of the hold-downs.

Figure 4.42 shows a variety of under-press conveyor configurations. The first and last configuration are typically used with a separate, sometimes slave-driven, horizontal conveyor that carries the parts out and drops them onto the inclined cleated belt conveyor.

Cleated belt conveyors are used for excavation work, to raise dirt up and dump it on top of a pile, or in farming to carry hay bales up to the haymow. They can be used to convey almost any loose, free-flowing material as well as other items up inclines that are typically too steep for normal flat belt conveyors.

4.6 TROUGHED BELT CONVEYORS

The last type of belt conveyor to be covered is troughed belt conveyors. Thousands of pages have been written on the engineering associated with troughed belt conveyors and bulk material handling. We look here at the basic design considerations and point out things to look out for and the pitfalls to avoid.

Troughed belt conveyors are used almost exclusively for bulk material handling. The materials carried range from coal and iron ore to dog biscuits and crackers. The primary point is that in a bulk application, there is no concern for product orientation. Troughed belt conveyors can be up to several hundred feet long, and unlike typical belt conveyors that require a straight, level installation, troughed belt conveyors follow the terrain over which they are installed (see Figure 4.43).

The two subsets of troughed belt conveyors are pipe conveyors and slider/roller (S/R) conveyors. Pipe conveyors are very similar to typical troughed belt conveyors

FIGURE 4.43 Troughed belt conveyor.

FIGURE 4.44 Overland troughed belt conveyor. Note that it can follow the general contour of the land.

and S/R conveyors are primarily used in glass manufacturing plants for carrying cullet, broken scrap glass.

Just like with flat belt conveyors, several standard sections make up a troughed belt conveyor. The three primary sections are the intermediate beds, end pulleys, and drives. Troughed conveyors can range in length from 10- to 20-ft. long up to thousands of feet long. These longer conveyors follow the terrain over which they are constructed as shown in Figure 4.44. It is not uncommon for these conveyors to turn corners and climb hills. The longest troughed belt conveyor is more than 9.5 miles long at an iron ore mine in KweKwe, Zimbabwe.

The intermediate bed is made up of a series of idlers on a structural steel frame with a return roller underneath. The idlers are typically a series of three equal-length rollers, one flat and the other two inclined up on either side. As the belt sits down on them, it is forced into a troughed shape (see Figure 4.45).

The idlers are available in a variety of angles with 20 degrees, 35 degrees, and 45 degrees being the most common. The size of the rollers and their construction can be changed for various applications. Figure 4.46 shows a typical cross-section of a conveyor with a 35-degree troughing idler. Lighter duty applications typically use lighter weight rollers as small as 1.9 in. in diameter. Conversely, heavy-duty applications use larger rollers, typically up to 7 in. in diameter. Different applications require different bearings in the roller as determined by the environment in which the conveyor is to

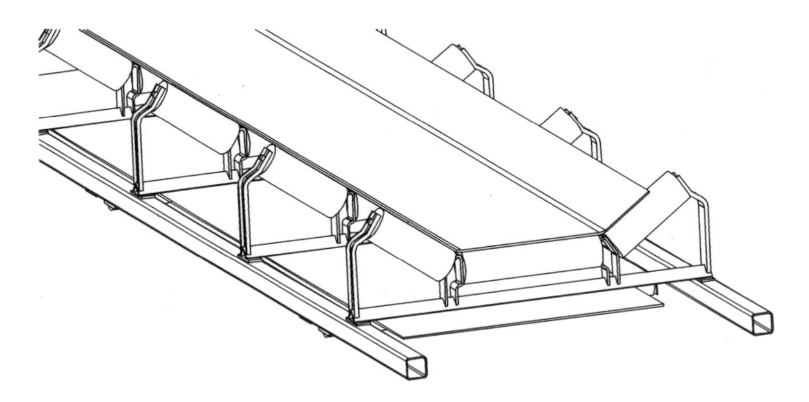

FIGURE 4.45 Simplified troughed belt conveyor design.

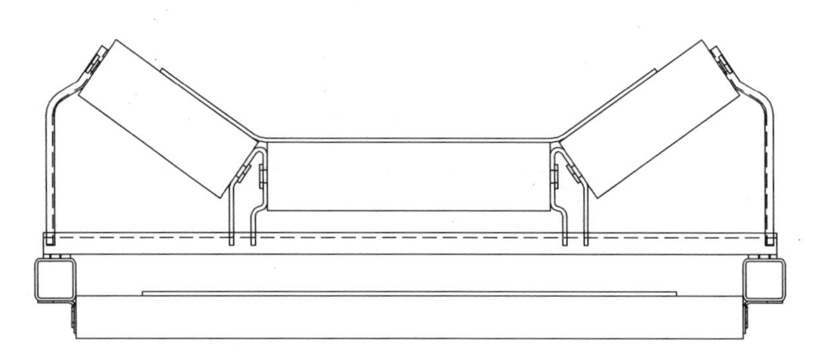

FIGURE 4.46 Cross-section of a troughed belt conveyor.

operate. If the conveyor is carrying dog biscuits, the bearings need to use food-grade grease. Likewise, if the conveyor is in a coal mine, the bearings must be well greased and have seals to keep out dust and debris.

Specialty idlers are made for certain areas of a conveyor. Some idlers are made with heavy rubber coatings that are used where heavy product such as aggregate is being loaded onto the belt. These are found under hoppers or chutes and are used to absorb the shock of the falling product. Some are designed to help remove product from the surface of the belt.

Skirt boards are used at load points to keep the product centered on the belt. They typically extend beyond the transfer point until the product is settled and moving at full conveyor speed.

The belts on troughed belt conveyors are tracked (made to run straight) in one of several ways. One way, similar to flat belt conveyors, is that the snub and return rollers are used to track the belt. By turning them, the belt can be directed back toward the center. This is most effective on shorter conveyors, typically less than 100-ft. long. Another popular way is through the use of a training idler, an idler that is mounted on a pivoting base that is equipped with guide shoes or rollers. As the belt pushes against one of these shoes or rollers, the idler pivots and steers the belt back to center. There are both troughed and return-style training idlers.

To aid in tracking the belt, it is important to make sure that all loading and transfer points drop the product onto the center of the belt. If product is loaded on one side, the belt will be pushed to the other side.

Interestingly, it was belt tracking that actually came up during a job interview. I was walking through the assembly building at Orteq when the Director of Operations mentioned to me that they were having a problem tracking a belt on a troughed belt conveyor. After glancing at it, I simply stated that as long as the pulleys of all the idler sets are square to the belt path, the only issue could be the lacing. I told him they should recut both ends of the belt as square as possible, relace it, and try again. He spoke to the assembly supervisor and we continued on. Only after I got the job offer did I find out that I was right. They recut and relaced the belt, it ran without issue.

Due to the far larger conveyor size, the drives are also far larger. The center drive, like the center drive of the flat belt conveyor, can be mounted anywhere along the length of the conveyor, although it is generally within the third closest to the discharge end. That being said, drives for troughed belt conveyors are almost exclusively end drives. The reason for this is fourfold. First, an end drive is less expensive. Second, to get the wrap needed around the pulley to maximize traction is easiest at the end. Third, it minimizes the amount of belt under tension. Finally, because when conveying product over great distances or out a strip mine, the electrical services for the motor are typically more readily available where the conveyor ends.

The drive pulley is typically rubber lagged to provide the traction needed to pull the belt. This lagging can be vulcanized to the pulley or can be replaceable. The replaceable lagging is more expensive but is very useful in locations where replacing the entire pulley is too difficult.

Tail or idler pulleys are typically winged or self-cleaning pulleys. Unlike the cylindrical shape that is usually thought of when talking about pulleys, the winged pulley looks like a multi-pointed star with the points blunted as shown in Figure 4.47. This allows anything stuck to the bottom of the belt to navigate the pulley without damaging the belt. Frequently as the belt flexes around the pulley the debris will come loosen and fall off. The center hub of the winged pulley is tapered so that any debris will fall out the ends of the pulley, thus the name self-cleaning. The herringbone winged pulley can be used for applications where straight-winged pulleys suffer too much wear. When handling very abrasive materials, the flat bars on the wings of the pulleys can be made from chromium carbide or AR steel alloys.

In typical flat belt applications, the product transfers smoothly off the end of one conveyor directly onto the end of the next. In this case, the diameter of the pulley

FIGURE 4.47 Winged pullies.

plays a critical role in how well the product makes the transition. However, in bulk applications, the product is dropped off of the end of one conveyor onto the next. Therefore, the size of the end pulley has little effect on how the product transfers. In flat belt conveyors, mid drives were used when the conveyor got too long and the drive pulley got too big for the product being conveyed to make a smooth transition. In troughed belt conveyors, the less expensive end drive is used most frequently because it really does not matter how big the drive pulley gets, as the product will still drop off the end.

The belt width is determined by the physical characteristics of the product to be conveyed. The lump size, the angle of repose, and the angle of surcharge for a particular product need to be known to properly determine the conveyor width. The recommended maximum lump size is summarized in Table 4.1.

The product to be conveyed is also an important factor in determining the belt material. If the product has sharp edges, a belt with a thicker top cover is required. Likewise, if lumps are large and heavy, the belt must be able to withstand repeated impact.

The belt of a troughed belt conveyor is frequently the single most expensive component and spare part. Therefore, it is important to make sure that the conveyor manufacturer is not overselling the belt. This does not mean that you should be stingy with the belt, because it is the heart of the conveyor. The conveyor manufacturer should be able to show you sound reasoning as to exactly why they chose a specific belt for your application.

As mentioned, unlike flat belt conveyors, troughed belt conveyors can navigate turns, both horizontal and vertical. In horizontal curves, the curve radius should be large enough to prevent the belt from tracking off of the idlers. The idler can be banked away from the center of the turn to aid in holding the belt in. Typically, the radius for horizontal curves varies between 1 and 2 km. Vertical curves have similar limitations. The concave curve at the bottom of the hill should be large enough to prevent the belt from lifting off of the idlers. At the top of the hill, the convex curve should be large enough to limit the loading on the idlers.

Finally, there are a variety of accessories that are made specifically for use on troughed belt conveyors. As mentioned previously, there are chutes, hoppers, and skirt boards. Belt cleaners, side idlers, and conveyor covers are also available.

Belt cleaners can be scraper blades that are pressed against the underside of the belt to scrape off any clinging product. Rotating brushes are also used to clean the belt. Several return idler designs exist that clean the belt as it goes by. Belt cleaners

TABLE 4.1
Lump Size versus Belt Width

Surcharge Angle	Composition	Width
20°	10% lumps/90% fines	1/3 belt
	100% lumps	1/5 belt
30°	10% lumps/90% fines	1/6 belt
	100% lumps	1/10 belt

are typically mounted directly behind the head pulley so that any material scraped off will fall with the rest of the material coming off the conveyor.

Side idlers are short, vertically mounted rollers that keep the belt centered on the conveyor. They can be equipped with a limit switch so that if a belt is too far out of track, an alert can be signaled.

In areas where material is being loaded onto the conveyor, it is good practice to do something to mitigate the impact of loading. This can be as simple as having the troughing idlers closer together in the impact area. Other solutions include impact idlers which are idler constructed of hard rubber rather than steel. Another solution is the use of impact beds. These are specially designed units that replace several idlers and consist of a series of bars that are parallel to the conveyor and have a rubber body to absorb the impact and a UHMW top to provide a low-friction surface for the belt.

Conveyor covers are used when the product being conveyed outdoors needs to be protected from wind and precipitation. This is very common when conveying grain, rock salt, or powdery materials that will be damaged by moisture or blown away by wind.

CASE STUDY – RAILCAR UNLOADER

While I was Director of Engineering for Orteq, we had to design a belt conveyor that could fit between the bottom of a hopper door on a railcar and the top of the railroad rail. The purpose was to unload frac sand from the railcar and transfer it to a pneumatic tank truck. The space between the railroad rail and the hopper door is a minimum 5″ [127 mm]. Additionally, we had to design the conveyor so that a hopper truck could drive over the conveyor to unload its contents. The final requirement was to keep track of the weight of the product being transferred so we didn't overload or underload the pneumatic trucks. The ultimate goal was to load the pneumatic trucks as fast as possible.

Due to the load being carried, we had to use a minimum 2-ply 220PIW belt. The lacing required for this size belt had a recommended minimum pulley diameter of 6″.

NOTE: As we discussed briefly in the Introduction, it is important to recognize ahead of time that when you break a rule, there is a price to be paid.

With only 5″ available and needing some clearance to move the conveyor in and out of place, we determined that the conveyor could be no more than 4″ thick. The 4″ thickness also worked well for allowing a truck to drive over it. That left us with only 3.5″ for the tail pulley diameter. This is significantly less than the 6″ minimum recommended by the belt lacing manufacturer. No one offers a 3.5″ diameter winged pulley. So, we designed and manufactured our own. We used a 1-½″ diameter shaft with six ¼″ × 1″ bars welded to it and then turned the weldment on a lathe to ensure it was perfectly round.

A standard cast steel 6″ diameter winged pulley should last for 5–10 years. Due to the pulley being undersized, we knew the pulley would be worn down by the hardened steel belt lacing and would have to be replaced approximately once a year. This was the price we had to pay for breaking the rules.

The horizontal section of the conveyor was reinforced to support the truck driving over it. We added side boards that could be raised hydraulicly once the conveyor was in position. This formed a trough for the belt to better contain the product being conveyed (Figures 4.48, 4.49).

To make the transition from horizontal to incline we had two 16" diameter turf tires being held down by gas springs.

Part way up the incline, we replaced three of the idler sets with new idlers mounted on a scale. We also added an encoder to measure the belt speed. We determined we could run the belt as fast as 1200 FPM using a hydraulic motor, but the scale lost its accuracy at any speed over 350 FPM. With the scale, encoder, and a small PLC, we could measure the weight of the sand moving over the scale and calculate the total weight of the sand that had been transferred from the railcar to the truck.

FIGURE 4.48 Railroad hopper car.

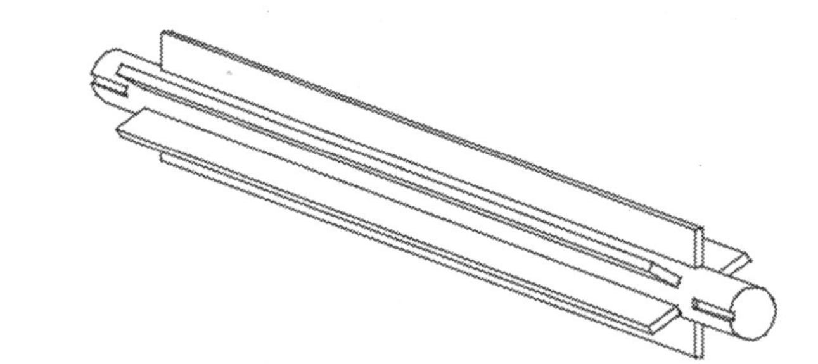

FIGURE 4.49 Welded winged pulley.

4.6.1 Pipe Conveyor

The primary difference between a typical troughed belt conveyor and a pipe conveyor is the geometry of the idlers. With a troughed belt conveyor, where product is loaded onto the belt, the geometry is in a troughed form. However, once the product is on the belt, the idlers get increasingly higher and tighter until the edges of the belt actually overlap and the idlers are on all sides, creating a pipe as shown in Figure 4.50.

Pipe conveyors are relatively new, compared to typical troughed belts that have been around for more than 150 years and have remained virtually unchanged since.

The loading zones, tail, and drive sections are virtually identical to a standard troughed belt conveyor; the primary difference is the intermediate section.

This pipe geometry offers some advantages. There is no need for covers because the belt creates the cover. Pipe conveyors can effectively navigate tighter vertical and horizontal curves with radii ranging from 100 m to 400 m. Finally, this type of conveyor can go up steeper inclines of up to 30 degrees rather than the typical maximum of 20 degrees for troughed belts.

Although pipe conveyors would seem an ideal configuration, they have a few drawbacks. The first is that they are slightly more expensive to build and maintain. Second, the pipe conveyor has a lower carrying capacity for a specific belt width. Finally, due to the pipe geometry, the conveyor profile is much taller than a typical troughed belt conveyor because the return belt is below the carry belt and unlike a troughed conveyor where the return is flat, the belt is in a pipe form again.

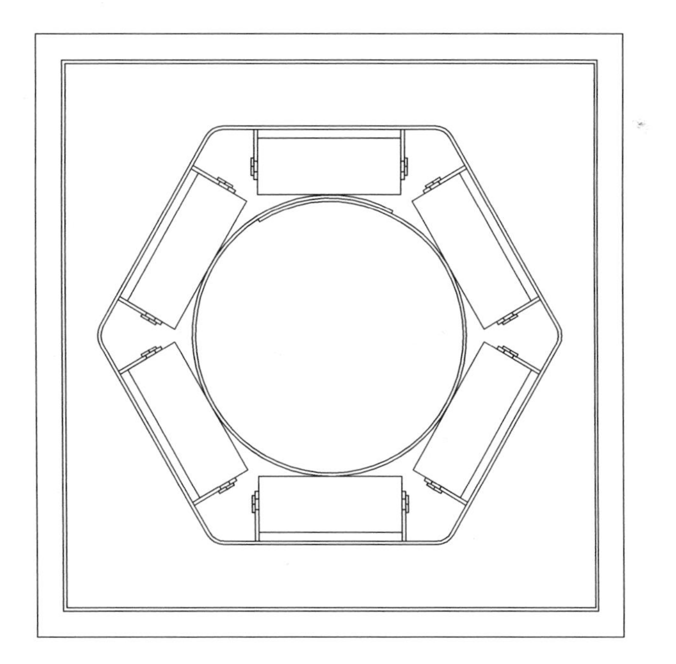

FIGURE 4.50 Cross-section of a pipe conveyor.

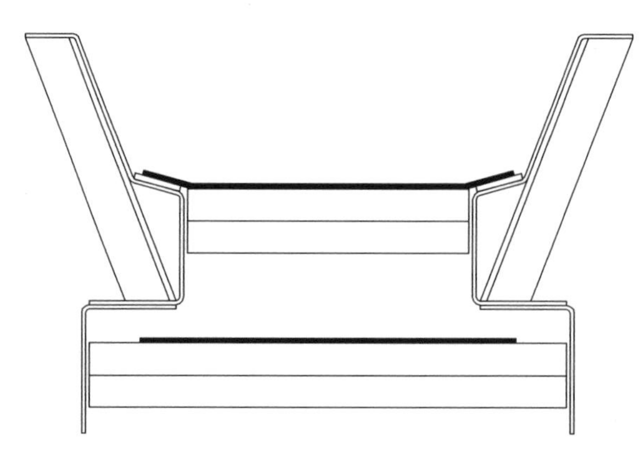

FIGURE 4.51 Cross-section of an S/R conveyor.

4.6.2 S/R Belt Conveyor

The S/R conveyor, as the name indicates, uses a combination of rollers and sliding surfaces to create the troughed belt geometry. S/R conveyors were developed for and are primarily used in the handling of cullet, or scrap and broken glass. For this reason, they are frequently referred to as cullet conveyors.

The primary differences between typical troughed belt conveyors and S/R conveyors are that the S/R conveyor uses only one roller to support the belt and two sliding surfaces that are integral to the side boards (see Figure 4.51).

4.7 BELT TYPES

Every belt manufacturer will, of course, tell you that they make the best belts. We are not going to attempt to settle that debate. As we have done with the conveyor itself, we discuss the various design and features of belting.

When you talk about conveyor belts, everyone's first thought is of a wide, flat, black strip of rubber. That is not completely inaccurate, but there is a lot more to that strip of rubber. A typical conveyor belt is made up of several sections: the top cover, the carcass, and the bottom cover as shown in Figure 4.52. Because we covered the full range of belt conveyors, we touch on the belts used in each of them here.

We will work our way down through the belt starting with the top cover. The cover can be made of urethane, PVC, Teflon, or synthetic rubber and can come in a variety of finishes and textures. PVC is the least expensive of the three materials, so it is the most popular for horizontal package conveyors. The drawback is that PVC will not work for many incline conveyors because it is too hard to grip the product being conveyed. PVC is unsuitable for bulk applications where the high impact from loading requires the flexibility of rubber.

On a typical horizontal flat belt conveyor and troughed belt, a smooth surface is sufficient. When moving up an incline or down a decline, to maintain a grip on the product, a textured surface is used. Most package conveyors use what is called rough-top belting; others will use belts with narrow longitudinal grooves (LG). The texture

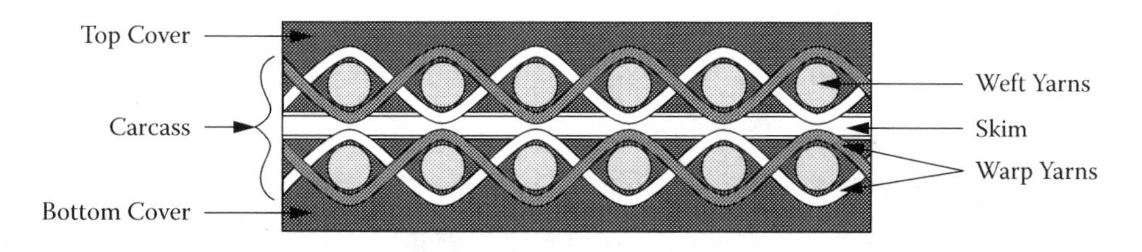

FIGURE 4.52 Belt construction.

is of an industry standard. For steeper inclines or for product that tends to be slipperier, there is another quasi-standard called diamond top. This is typically made of urethane because it is a softer and therefore more "grippy" material.

When working with sorters or any application in which the product must move side to side on the belt, a brushed surface is frequently used. This is created during the manufacturing process by removing any material from the carcass and providing a slicker surface. Other belts are specifically designed with an even slipperier surface.

The carcass is the part of the belt that actually does the bulk of the work. The carcass has four primary purposes: provide the strength required to move the loaded belt, absorb impact from material being loaded, provide load support, and provide fastener holding. The carcass can be made up of fabric, typically cotton, polyester nylon, or Kevlar®, but metal cords or wire mesh can be added for more pull strength.

The primary terms used when talking about fabric are warp and weft. The easy way to remember the difference is the weft runs from left to right and the warp runs the length of the belt, as illustrated in Figure 4.23. On flat, straight-running belt conveyors, it is preferable that the belt be stiff across the width of the conveyor and the edges do not curl up to keep product from getting under the belt. These belts utilize a monofilament weft and multifilament warp. The monofilament is much stiffer than the multifilament yarn. For troughed belt conveyors, multifilament yarn is used for both the warp and weft. This is sometimes referred to as a balanced weave, and this is also very useful for curved belts. When a curved belt is cut from a roll of belting, the resulting piece has the warp and weft pointing in different orientations throughout the arc of the belt (see Figure 4.24). With the balanced weave, the stretch is more consistent throughout the belt.

The carcass can be made up of multiple layers of fabric; these layers are called plies. A layer of bonding material between the plies is called the skim. The material used for the skim is typically the same as the top or bottom cover. The skim layer is an important aspect of the belt's strength, impact resistance, and flexibility.

When the belt gets longer, as with troughed belt conveyors, the fabric of the carcass can be reinforced with steel cords or wire mesh. The strength of the belt is referred to by its rated maximum recommended operating tension in newtons per millimeters width (NPM) or pounds per inch width (PIW). Most fabrics range from 25 PIW (4.5 NPM) to 300 PIW (52.5 NPM). As multiple plies are used, the strength of each ply is added together. Steel cord reinforced belt ranges from 600 PIW (103 NPM) to 5,000 PIW (876 NPM).

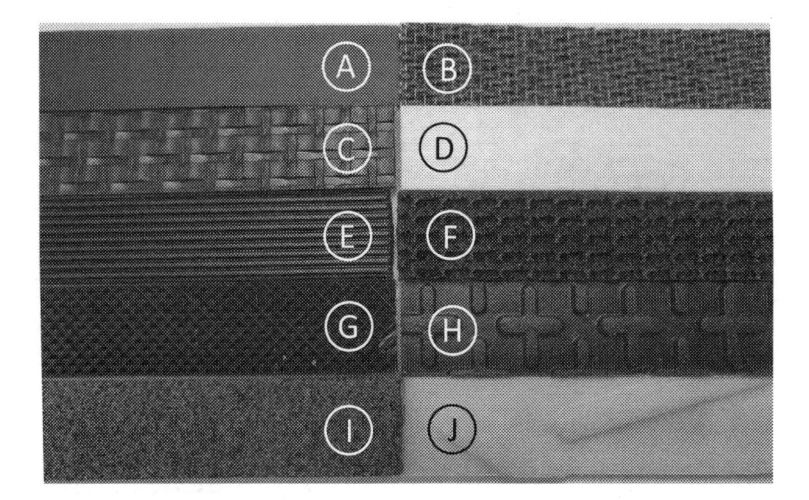

FIGURE 4.53 Belt surfaces.

A – Smooth Top, B – Brushed, C – Basket Weave, D – Smooth (Washdown), E – Longitudinal Groove (LG), F – Rough Top, G – Inverted Pyramid, H – Cross Boss, I – Needle-stitched, and J – Herringbone Boss

The bottom cover is the third part of a belt. On the vast majority of flat conveyors, especially where slider beds are used, the bottom has what is called a brushed surface. In this case, during the manufacturing process, the bottom cover is applied and then brushed back off so that the bottom ply is saturated, but this is minimal material left behind.

Depending on the application, the belt can be made with the sides sealed so that no moisture can be absorbed by an exposed edge of the carcass. This is true for all food applications, as well as many troughed belts, because of their frequent exposure to moisture. These belts are manufactured to a finished width, unlike typical belts, which are manufactured in 6-ft. (2 m) to 14-ft. (4 m) wide rolls and then cut down to the required width.

A variety of specialty belts are on the market, but we do not cover all of them here. One such belt material that does not fit the top cover–carcass–bottom cover scenario is called needle-stitched, which is a felt-like material that is made of random fiber and a binding material. This belt does not withstand high tension, but it does allow product to slide over it easily (Figure 4.53).

4.8 QUESTIONS

1. Why are monofilament belts preferred for most package conveyor applications?
2. A nose-over is not always required at the top of an incline belt conveyor. What are some reasons why not?
3. What are some reasons to use covers on a troughed belt conveyor?

4. True or False: A monofilament belt would work well for a pipe conveyor.
5. What is the primary reason a center drive should be in the third of the conveyor toward the discharge end?
6. Figure 4.17 shows a diagram of a box on an incline. What would limit the angle of the incline?
   a. The center of gravity is too far back.
   b. The belt cannot hold the box from slipping.
   c. Both a and b.
   d. Neither a nor b.
7. True or False: The outside radius of a spiral belt turn has the steepest angle.
8. What is the correct speed for a brake belt if the average product length is 18" and the required rate is 30 cartons per minute?
9. Which of the following does not contribute to determining pulley diameter?
   a. Belt material
   b. Lacing type
   c. Belt speed
   d. Belt width
10. Which belt in Figure 4.53 would be recommended for incline belt conveyors handling packages?
    a. A and B
    b. E and F
    c. A and I
    d. C and H

5 Static (Gravity) Conveyors

Static conveyors, or gravity conveyors as they are frequently referred to, are so named because there is no power source to keep the product moving other than human force or, if the conveyor is mounted on an angle, using Earth's gravity to keep the product moving. Wheel and roller are the two primary types of static conveyors. Each is discussed individually here.

Uninhibited access to gravity conveyors is required. As hang-ups and jams can occur on gravity conveyors, they should be applied only in applications where personnel can access the conveyor to clear jams. Gravity conveyors should not be used in automated systems where complete control of product flow is critical.

Wheels are better than rollers for lightweight loads. Empty cartons and lightweight loads travel better on wheels than on rollers because the force required to overcome the inertia to start several wheels is far less than that required to start a single roller.

Rollers are better with steep declines and low rates. If the carton flow rate is low, roller conveyors can be used in steeper applications than wheels without excessively accelerating the cartons because the roller mass acts as a retarder. In high-rate applications, carton speed will increase due to the turning momentum of the rollers.

5.1 STATIC WHEEL

Static wheel conveyors are also referred to as skate wheel conveyors because the primary component was originally based on old-fashioned steel roller skate wheels. Although there is much controversy over their history, it has been reported that the first skate wheel conveyors were developed during prohibition to assist in removing equipment from closed breweries.

The wheels can be mounted either high or low in the frame (see Figure 5.1). High-mounted wheels are set so that the top of the wheel is above the side frames. Low-mounted wheels are set down between the side frames.

Static wheel conveyors are typically available with frames constructed of steel or aluminum. The wheels and spacers are also available in either material. Wheels are also available with a variety of plastic or rubber coatings or tires and can also be made of some plastics. The aluminum frames and wheels make a lightweight conveyor for portable applications, but the aluminum frame does not have the same weight capacity as a steel frame. In all cases, it is important to ensure that the frame can support the load.

Two dimensions are especially important when working with static wheel conveyors. The first, width, is the most obvious. Width is a function of the width of the products to be carried. If the wheels are set low, the conveyor should be wider than the widest package.

DOI: 10.1201/9781003376613-5

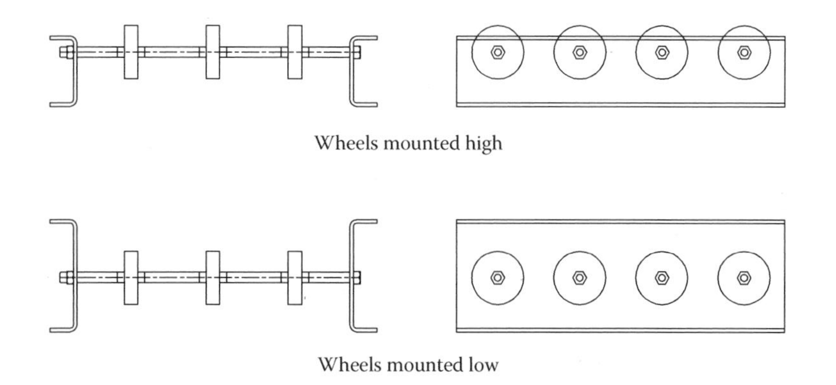

FIGURE 5.1 Skate wheel conveyor.

If the wheels are set high, it is not uncommon to see products overhanging the sides of a static conveyor as long as it is not feeding subsequently narrower conveyors. For this to be possible, the conveyor route must be free of obstructions and the product must be firm enough to prevent overhang below the plane of the conveying wheels. Use caution in allowing product overhang through a curve. Typically, the product can be 25 percent wider than the conveyor.

The second dimension is the wheels per foot. This is a function of the smallest product to be handled, in both length and width. Each row of wheels is mounted on a common axle that is typically ¼ in. in diameter. The axles are most frequently mounted on 3-in. centers; however, they can be on centers as small as 1½ in. or as large as 6 in. The axle spacing is based on the length of the product. The axle spacing should not be more than one-third the length of the shortest product to be handled. The side-to-side spacing of the wheels or the number of wheels on each axle is determined by the width of the smallest package as well as the maximum load to be carried.

Curve frames are available to match the construction of the adjoining straight sections. The radius of the curve is determined by the product to be conveyed. The length and width dimensions of the product to be handled determine the width (between rails) of the curve and, generally, the width of the curve determines the width of the adjacent connecting straight conveyor. If a package is particularly long, similar to a box of fluorescent light tubes, the adjoining straight conveyor would be too wide for such a product. This condition creates a costly conveyor. This situation can be corrected by using a narrower width curve and offsetting the guard rails to permit the product to overhang the curve but still keep the adjacent straight conveyor to a reasonable width and help keep the cost down (see Figure 5.2)

$$BF = \sqrt{(Radius + PackageWidth)^2 + \frac{PackageLength^2}{2}} - Radius + 1$$

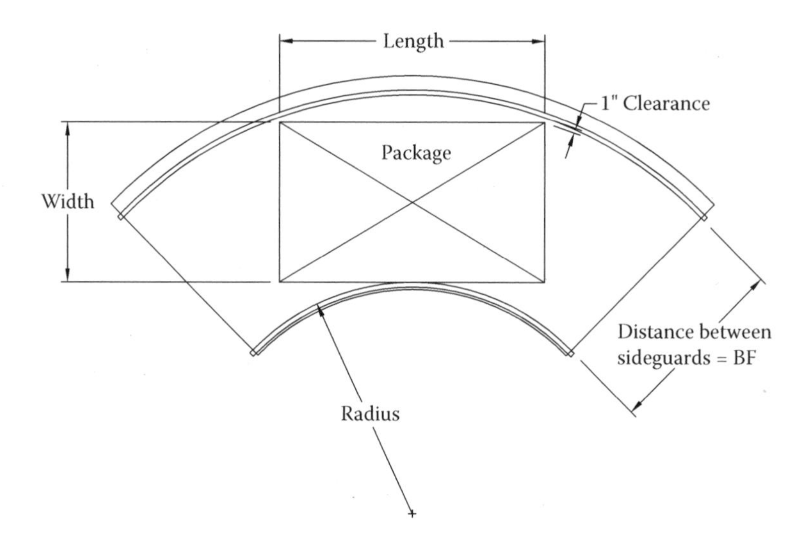

FIGURE 5.2 Curve sizing based on product size.

5.2 STATIC ROLLER

The static roller conveyor follows the same concepts as static wheel conveyor. The rules for sizing the conveyor still hold true, except we are not concerned with wheel spacing. Again, frames are available in both steel and aluminum, and there are far more roller options.

The roller sizes range from ¾ in. up to 3½ in. in diameter. The most popular sizes are 1.4, 1.9, and 2.5 in. in diameter. Typically, the rollers have a steel shell or body, but smaller diameters are available in both steel and aluminum shells. All of the sizes are available with various plastic sleeves or coverings as well as being made of plastic.

A static roller conveyor is made up of a series of rollers on 2-, 3-, 4-, or 6-in. centers mounted between side frames. Like the skate wheel conveyors, the rollers can be mounted high or low in the frame. Curved sections are available with the same rollers as the straight conveyor or tapered rollers as shown in Figure 5.3. The straight rollers are lighter and less expensive; they also turn easier when dealing with lightweight products. The drawback is that the product is forced to the outside of the curve, creating drag against the guard rail. A curve with straight rollers will not maintain the product's orientation well. If the product is round, then the orientation issue is not a factor. Double rollers (curves with a center frame that runs through the center of the curve that splits the roller length) will convey a square or rectangular package better than the single straight roller curve because the outer lane of rollers can turn faster. Tapered rollers are typically preferred because the product orientation is easily maintained. However, tapered rollers are heavier and will not operate satisfactorily with very light products.

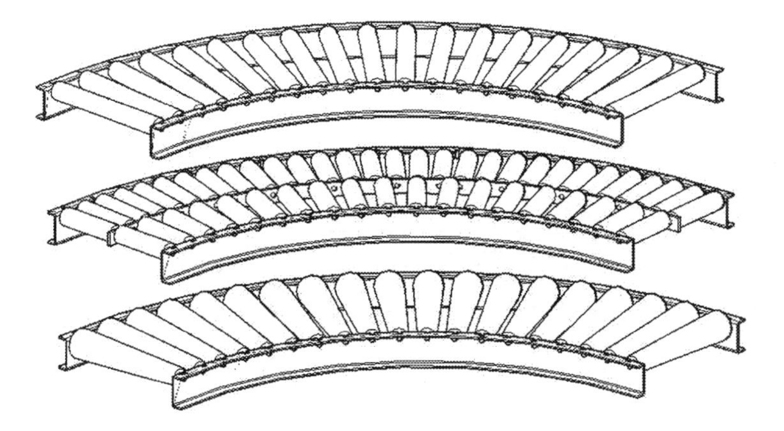

FIGURE 5.3 Roller curves.

5.3 APPLICATION DETAILS

Probably no other type of conveyor is applied as often as static wheel or roller conveyors. They can handle various packaged materials efficiently for distances as short as 2 ft. or as long as 100 ft. or more. Any item from lightweight widgets to sheets of plywood to cement blocks can be moved on static wheel or roller conveyors.

Static wheel conveyors, as a rule, are generally used for handling smooth-bottomed, wood, fiber, corrugated, or plastic containers. Semi-rigid, smooth-bottomed bags, or bales can also be conveyed on some wheel conveyors. Static wheel conveyors are not recommended for handling cans or drums with chimed bottoms, open-bottomed crates, or damp, soft cartons. Each wheel is a support point and can create an indentation in the bottom of a cardboard box, so it is important to limit the point loading without adding so many wheels that the costs are too high. This point-loading aspect becomes very important when dealing with bags or bundles. If the product is too moldable, it will be unconveyable. If there is any doubt, the product and container should be tested on an appropriate wheel conveyor to help determine the suitable pitch for the conveyor.

The static roller conveyor is also used for handling smooth-bottomed wood, fiber, corrugated, or plastic containers but is also effective in conveying crates; cans or drums with chimed bottoms; kegs; and long, narrow packaged materials.

As you will notice, a wider variety of package types can be carried on roller conveyors, but wheel conveyors can be substituted where a portable gravity conveyor is required as long as the product will convey properly.

If using these nonpowered types of conveyors, it is important to look at the angle or pitch at which the conveyor is to be installed. On average, the pitch required for corrugated cartons is 6 in. per 10-ft. straight section, 5 in. per 90-degree curve. Products with hard or firm bottoms require less pitch, and soft-bottomed products require more pitch. Pitch requirements decrease as conveyable product weight increases. Experience and testing are the only practical methods for determining actual pitch requirements.

The pitch required will vary depending on whether the wheels or roller bearings are dry or grease packed and on the ambient temperature and humidity. The angle of decline is also application specific. In the case of static roller conveyors, the first package will start the roller spinning. The second package, if it follows close behind, will not expend some of the gravitational forces working on it to overcome the inertia of the rollers and would travel a bit faster than the first package, and the third package would travel even faster. Over a span of time, say a day's production, trains of product could be traveling very fast on a conveyor originally pitched to start package movement from rest. Also, the excessive speed characteristics of longer distances can damage products as they come crashing to a stop, causing product jackknifing and overrun of curves or spurs. For gravity lines longer than 20 ft. where product can accumulate, review the application for high line pressure and impact. Whenever longer gravity runs are required, carefully analyze these concerns. When dealing with wheel conveyors and certain small-diameter roller conveyors, it is possible to tighten the axles to create drag on a group of wheels or rollers to slow the product down. Retarding devices such as inertia wheels are also available to help control this situation for gravity roller conveyors.

Gravity curves are not recommended for accumulation of anything except round or cylindrical products. The line pressure will prevent rectangular packages from maneuvering the gravity curves and generally will force the packages against the outer guard, thus blocking the free flow of further packages.

One of the specialized applications of static rollers is in storage systems. Distribution systems frequently employ pick modules where product is stored in a heavy-duty rack or what is referred to as a flow rack. The rack is equipped with a static pin roller conveyor mounted at an angle so that as product is removed by the operator, the remaining product flows down to fill the opening.

Some ergonomic details require attention when applying static wheel or roller conveyors. First, as previously mentioned, the elevation at which operators will load and unload the conveyor has a direct impact on the pitch of the conveyor and therefore how well product will flow down the line.

Second, as product flows down the conveyor it will impact product that has stopped ahead of it. This impact can cause injury to an operator who is trying to remove product from the conveyor. Operators should always handle product from the sides and never allow their hands and fingers to get between products.

5.4 SPURS AND "Y" CURVE SWITCHES

Gravity roller or wheel conveyor spurs are used for merging and diverging of packages onto or off of a main-line transportation conveyor (see Figure 5.4). The standard angles are 30 degrees, 45 degrees, and 90 degrees. The 45-degree diverging unit is not normally recommended for automatic diverging of packages but should be manually attended. Switches utilizing both fixed and pivoting skate wheels provide a simple method of diverting or converging products from one line to another. The diverging units can be manually actuated or automatically actuated using an air cylinder or small motor.

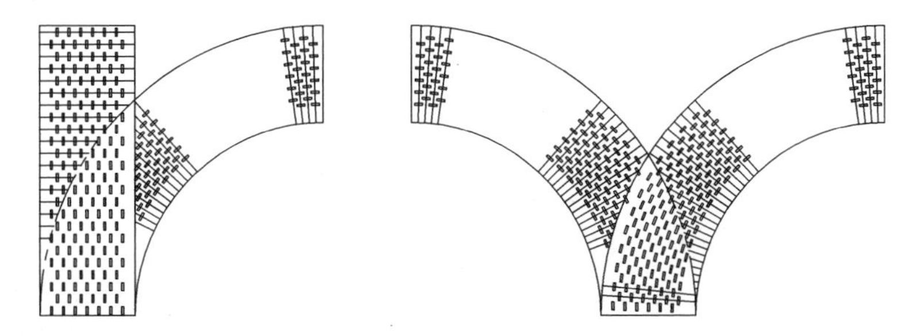

FIGURE 5.4 Spurs and "Y" curve switches.

5.5 GATE SECTIONS

Figure 5.5 shows an example of a hinged section used as a gate; these are available for vertical movement of the section to provide access for personnel, lift trucks, or other equipment. Heavier gate sections can be equipped with counterbalances that can be either heavy-duty die springs or gas cylinders. Some manufacturers offer horizontal gate sections with a pivot pin on one end and caster supports on the opposite end. This is used primarily where the conveyor to be opened is quite long and would be unwieldy to hinge vertically.

5.6 FLEXIBLE CONVEYORS

Flexible static conveyors are an available specialty item. They offer, as per their name, flexibility in a layout. They can bend around corners without using traditional rigid frames. This type of equipment is handy for loading and unloading trucks as well as for temporary applications.

Flexible conveyors are available with both wheels and rollers (see Figure 5.5a, b). The wheeled version provides better product tracking than the roller version; the product will better maintain its orientation relative to the conveyor. However, it has the same product conveyability limitations as the static wheel discussed previously. One aspect to consider with flexible conveyors is that as the conveyor is stretched

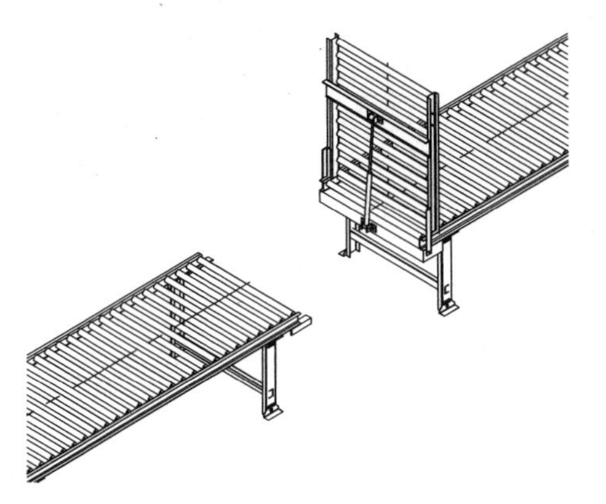

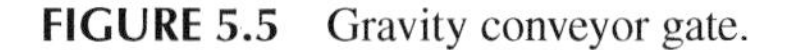

FIGURE 5.5 Gravity conveyor gate.

out, the axle or roller centers begin to grow. Expansion limiters are available to restrict the expanded axle centers to a length appropriate for the packages being conveyed. However, these expansion limits also affect the expanded to compacted length ratio of the conveyor.

5.7 EXTENDABLE GRAVITY CONVEYOR

Extendable gravity conveyors are used for truck loading. For loading, the conveyor can be extended into the front of the truck. As loading progresses, the conveyor is gradually pushed back (retracted). This conveyor is not recommended for unloading a truck. The conveyor can be moved quickly and easily on casters that can be locked into position (see Figure 5.6). Some manufacturers offer powered drive wheels to aid in extending and retracting the conveyor (Figure 5.7).

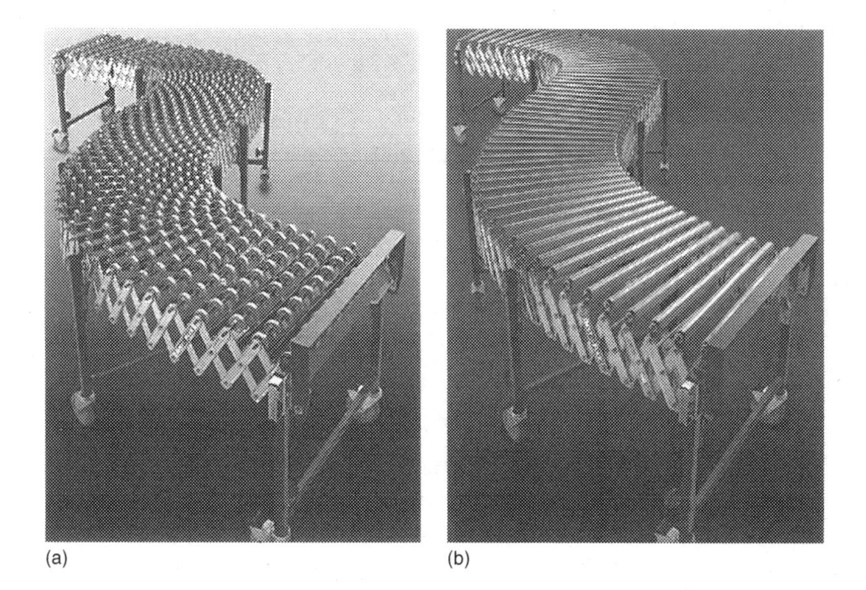

FIGURE 5.6 (a) Flexible skate wheel conveyors and (b) flexible roller conveyor.

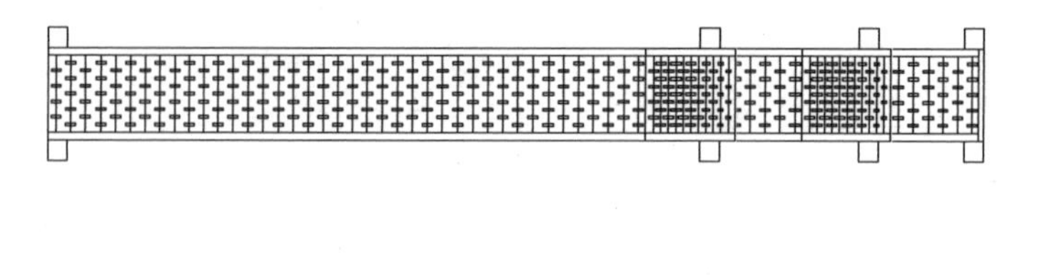

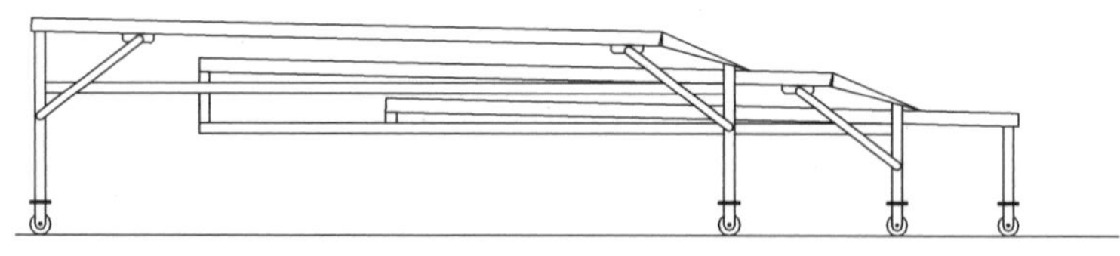

FIGURE 5.7 Gravity extendable conveyor.

5.8 FLOW RACK

As mentioned earlier, various types of static wheel and roller conveyors can be used in flow racks. Flow racks are used in distribution centers for picking operations as well as for parts storage in assembly line operations. Depending on their design, flow racks can handle cartons as well as pallets. Figure 5.8 shows some typical flow racks with (a) a reclined front, (b) a straight front, and (c) a pallet flow rack. Both carton and pallet flow racks can utilize either static wheels or rollers. Obviously, the pallet flow uses heavier wheels or rollers to support the loads.

Many manufacturers have specially designed rails with wheels attached for carton flow rack. Caution must be used when selecting a wheeled flow rail to ensure that it is of sufficient structural strength to support the product.

One of the nice features of wheeled carton flow is that there is nothing limiting the width of the boxes that can go on it. Roller carton flow works best with cartons that are no wider than the individual carton flow lanes. With the wheeled carton flow, there is nothing that prevents a box from straddling multiple lanes. Each manufacturer offers a variety of options to enhance a carton flow installation.

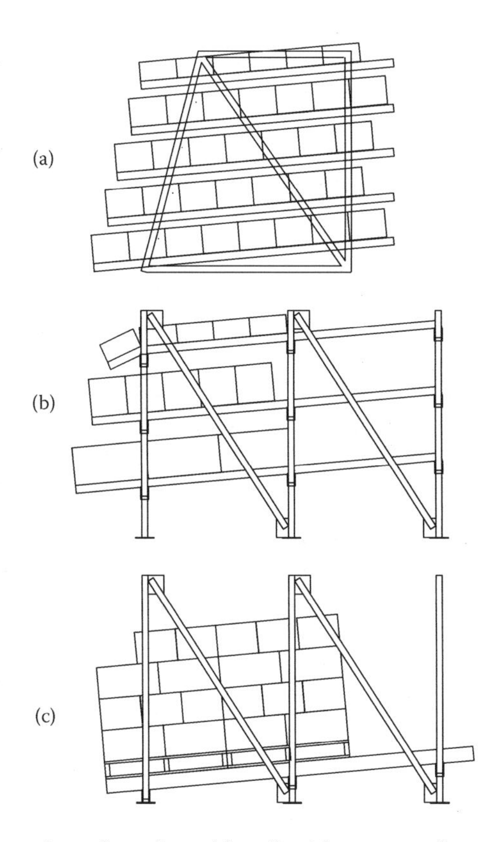

FIGURE 5.8 Flow rack configurations: (a) reclined front carton flow rack, (b) straight front carton flow rack, and (c) pallet flow rack.

For carton applications, there are several choices available:

1. Will product be removed from the open carton (split-case picking)? If so, adequate space must be provided to remove the product from the top of the case. This can be done by using a sloped front rack, using staggered levels, or using a pick tray or other methods to tilt the box out 15 degrees to 30 degrees. A combination of these can be used with the pick tray on the upper levels and the staggered levels below. Notice the tilted box is in the preferred position for better visibility and access. Also, it is important for the person picking to be able to see in the box.
2. Will full boxes be removed from the rack (full-case picking)? The spacing between levels of the flow rack must allow not only space for the carton to move along the conveyor but also sufficient space for the carton to be lifted over the end stop. This application typically utilizes the straight front rack. For ergonomic reasons, always keep in mind the weight of the cartons that will be stored on the top level or the bottom level and the typical height and strength of the people using the rack.

When dealing with pallet flow, it is important to consider how the send pallet will affect the operator that is removing the empty pallet. Flow retarders may be required to keep the second pallet from moving forward too quickly when the empty pallet is removed.

In designing the flow rack area, it is important to consider how the empty cartons or pallets will be disposed of. Will there be trash receptacles or an empty carton conveyor? Many pallet flow layouts include a row where empty pallets can be stacked. The rails are angled back out of the picking area so the empty pallet can be picked up and removed.

5.9 QUESTIONS

1. What would be a major detractor in using an extendable gravity conveyor for trailer unloading?

2. If the conveyor is 18” between the frames with the wheels set low, what is the longest 15” wide box that will move freely around the curve that has a 36” inside radius?

3. Which of these are a drawback of skate wheel conveyor?
   a. Even with only a slight decline product may move easily.
   b. Limited in available width due to the small axle diameter.
   c. The bottom of softer products may form to the wheels and need to be pushed manually.
   d. All of the above.

4. What are the advantages of a skate wheel curve over a straight roller curve?

5. True or False: Boxes on a flexible roller conveyor maintain their orientation around turns.

6. Which of these is not a consideration when designing a carton flow system?
   a. Disposing of the empty cartons
   b. Carton heights
   c. Operator visibility of product in the cartons
   d. Width of individual wheeled lanes
7. What is the primary reason gravity conveyors are not ceiling hung for overhead applications?
   a. Hanging a conveyor is too expensive an installation for an inexpensive conveyor.
   b. Hanging gravity conveyors are difficult to access in case of jams.
   c. If the ceiling sags due to conveyor loading, the pitch of the gravity conveyor may change.
   d. Gravity conveyors make too much noise for overhead uses.
8. Which of the following is not a problem to consider with long runs of gravity conveyors?
   a. Heavy product moving too fast and damaging lighter product when it stops.
   b. Hard plastic totes moving too fast versus soft cardboard boxes.
   c. Totes bridging up as line pressure builds.
   d. The floor supports are tall enough to support the infeed end.

6 Powered Conveyors

There is a wide variety of powered conveyors beyond the basic belt conveyors already covered. We have separated powered conveyors into three main groups: live roller transportation, live roller accumulation, modular conveyors and sorters. As you might have guessed from the grouping, the vast majority of powered conveyors, beyond basic belt conveyors, are roller conveyors.

Many unique powered conveyors have been built for specific applications. To cover all of these would require multiple volumes, and once they were complete, something new still would not be covered. For instance, one conveyor we built used two 50-mm wide, steel-reinforced urethane timing belts to convey large rectangular glass funnels. To create accumulation zones, we used urethane-padded, miniature, air-operated scissor lifts to lift the funnels off of the belts. To top it off, the entire conveyor was constructed of 304 stainless steel. This conveyor is useful for only a very small group of products. As you can see, new and unique conveyors are being developed all of the time. Development of these specialty conveyors is quite costly and should be undertaken only when all standard conveyor options have been exhausted.

6.1 LIVE ROLLER CONVEYORS

Rollers, of course, are the heart of roller conveyors. The rollers that are used are typically 50 mm (1.9 in.) in diameter. The rollers' shell, the outer surface that you see, is made up of round tubing with a wall thickness of 16–12 gauge. The heavier material is used for wider conveyors or those carrying heavier loads. Depending on the application, the roller shell can be plain steel, galvanized, or even zinc or chrome plated. The rollers are usually supplied with a spring-loaded $^{7}/_{16}$-in. hex shaft. Several variations on this have been developed to minimize the noise generated by the rollers rattling in the conveyor frame. For handling smaller products, some manufacturers offer live roller conveyors with 35-mm (1.4-in.) diameter rollers.

Depending on the application, the rollers can be coated with a slide-on urethane or Teflon sleeve or a sprayed-on or dipped coating. Coated rollers are used to either pad the hard steel surface or to increase the friction between the roller surface and the product being conveyed.

Keep in mind that the application might require special rollers. For instance, we designed a conveyor for handling rolls of photographic paper. To avoid bruising the edge of the paper, the rollers had to be much straighter than typical conveyor rollers. All rollers are slightly banana shaped to a certain extent. No roller is perfectly straight; they all have some amount of bend to them, typically a straightness or total indicator runout (TIR) of 1.14 mm (0.045 in.) for a 560-mm (22-in.) long roller. With the rolls of paper, if the roller that the roll of paper was on happens to be arched down and the next one is arched up, the leading edge of the roll would, ever so slightly, stub

DOI: 10.1201/9781003376613-6

into the next roller. This would bruise the paper and create cloudy areas in the subsequently developed pictures. Therefore, we used rollers that had a TIR of only 0.25 mm (0.010 in.). Additionally, the rollers had to be stainless steel rather than steel due to reactivity of the paper coating. This presented an interesting challenge.

Another customer had rollers carrying glass trays, so they requested coated rollers so that the glass would not hit the steel rollers. When we tested the system with empty trays, it performed very well. Typically, a system is tested with the entire breadth of product, but in this case the customer's product that would be carried on the trays weighed less than half a pound and was highly secretive, so testing with just the tray was determined to be sufficient. What the customer was not aware of was this inherent bend in the rollers. It was not until they were testing with product on the trays that they realized that the unevenness of the rollers would cause the product to vibrate and even fall off of the trays. Because we had fulfilled our contract based on their specifications, at considerable expense, they subsequently had to sort through and replace any rollers that had too much bend in them. This negatively affected their budget and their schedule. This illustrates the importance of knowing all aspects of the handling requirements and communicating them to the conveyor supplier.

When you enter a facility with several roller conveyors, one of the primary sounds you hear is the ring of the rollers. This is caused primarily by the semi-precision bearings used in most rollers. Much development effort has been expended in an effort to offer quieter, longer-lasting bearings for the rollers. Typically, these quieter bearings are more expensive, so they are reserved for higher-speed applications, typically over 1 m/sec (200 FPM). The noise generated by the less expensive bearings is tolerable at lower speeds. If noise is a concern, the quieter bearings can be used at lower speeds as well.

6.1.1 Live Roller Transportation

6.1.1.1 Belt-Driven Live Roller

The belt-driven live roller (BDLR) is the simplest of the various types of live roller. BDLRs use a bed of rollers as the conveying surface and a belt underneath to drive the rollers.

BDLR conveyors can be driven by a variety of belt shapes. The belt can be a flat belt like a typical belt conveyor, but it can also be a round belt, a v-belt, or a narrow flat belt 25-mm to 75-mm (1-in. to 3-in.) wide. The flat belt version is used for longer conveyor applications and the round or v-belts are used for shorter conveyors. The flat belt can only drive straight conveyors. The round or v-belt can be used to drive a combination of straight and curved conveyor beds.

Like a belt conveyor, there are several common sections of a BDLR conveyor. The drive on the flat belt version, because it is pulling a belt, is virtually identical to a flat belt conveyor drive. The vast majority of flat belt BDLRs utilize a center drive. The center drive not only offers a more positive pull but also includes a take-up for belt tensioning. As shown in Figure 6.1, the center drive is virtually identical to that of a flat belt conveyor. Some manufacturers actually do use the exact same drive assembly for both BDLR and flat belt conveyors. End drives, although less

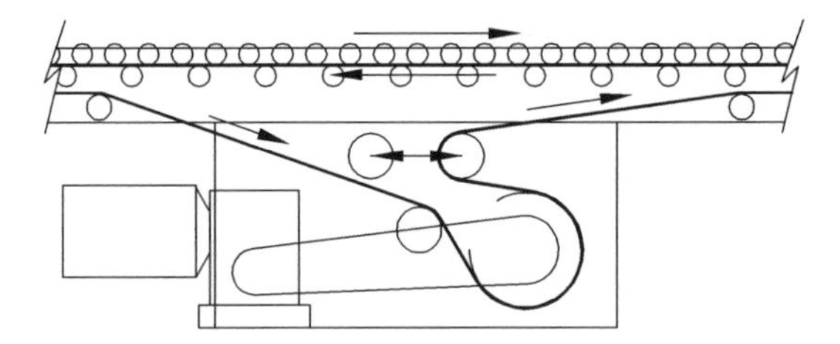

FIGURE 6.1 Center drive.

FIGURE 6.2 Flat belt-driven live roller.

expensive, do not have adequate take-up so the end pulley must be used and the take-up available there is minimal. Therefore, end drives are used only on fairly short BDLR conveyors.

The drive of the round or v-belt versions is a very simple driven sheave. On all but the shortest conveyors, a separate take-up, typically spring-loaded, is provided to keep the belt tensioned.

Keep in mind that the belt is running in the opposite direction of the product being conveyed.

Intermediate beds are typically constructed of two formed channel side frames with the carry rollers exposed on top and a second set of pressure rollers directly below that are on wider centers (see Figure 6.2). These hold the belt against the bottom of the carry rollers. The carry rollers usually have 2-, 3- or 4-in. centers and the pressure rollers have 6-, 8-, or 12-in. centers. There is a third set of rollers that support the return belt, and these are spaced much farther apart.

With the round or v-belt versions, the pressure and return rollers are replaced by small sheaves (see Figure 6.3). While the flat belt typically runs beneath the center of the rollers, the round and v-belt are typically justified along one side or the other.

To adjust the drive pressure supplied by the belt, most manufacturers offer a method to adjust the support pressure supplied by the pressure rollers or sheaves. Some are individually adjustable, and others use a steel angle or bar to adjust a series

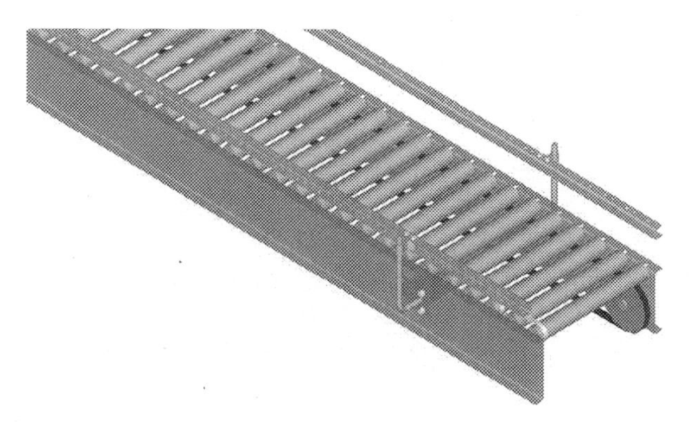

FIGURE 6.3 V-belt-driven live roller with end sheave.

of rollers at a time. This allows the pressure to be reduced where product can accumulate occasionally or increase in areas of merges or slight inclines. Inclines and declines are possible with BDLR but should be limited to no more than 5 degrees.

Instead of end assemblies as with belt conveyors, BDLR conveyors typically have what are referred to as end beds. These beds have idler pulleys and snub rollers built into the end of them. Due to the average 4- to 6-in. diameter pulleys or sheaves that are used, the last couple of rollers do not come in contact with the drive belt. To avoid a dead zone in the rollers, these otherwise undriven rollers are usually slave driven by one or two of the last belt-driven rollers. The slave drive can be done using grooved rollers and round urethane bands (similar to the lineshaft conveyor discussed later) or by using thin 12- to 25-mm (½- to 1-in.) wide flat urethane belts as shown in Figure 6.4.

Curves and merge beds can also be incorporated into the BDLR conveyors that use round or v-belts. These sections can be stand-alone conveyor units as well. Merge beds have the belt along the long side in an effort to drive as many rollers as possible.

Curves usually have the belt under the inside of the curve if it is used as a stand-alone unit. When the curve can be part of a longer conveyor, in an effort to keep the

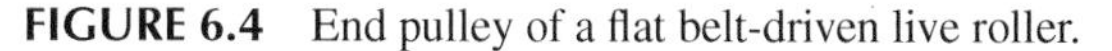

FIGURE 6.4 End pulley of a flat belt-driven live roller.

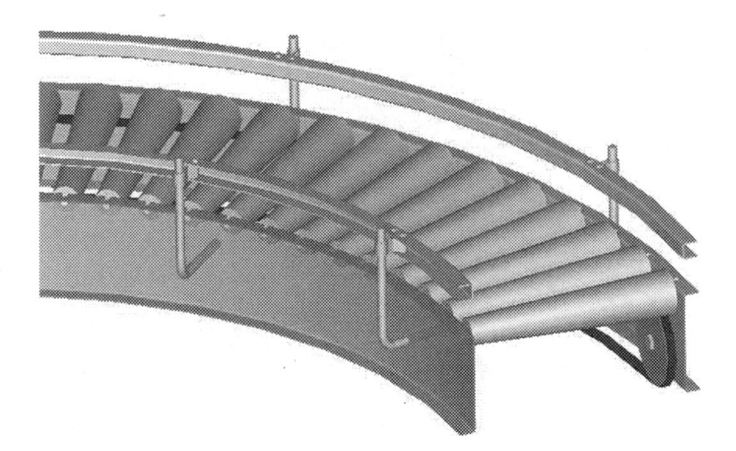

FIGURE 6.5 V-belt-driven live roller curve.

rollers running at a more consistent speed, the belt is under the outside end of the rollers (see Figure 6.5). Curves are typically constructed using tapered rollers. The rollers are bigger toward the outside of the curve. This differential in roller size allows the product to maintain orientation with respect to its direction of travel.

When a conveyor incorporates multiple curves, merges, or both that require the belt on opposite sides of the conveyor, a jackshaft is employed to transfer the drive force from the belt on one side to a belt on the other side.

Some manufacturers offer methods to slave drive the shorter round and v-belt-driven conveyors from the stronger flat belt BDLR. Slave driving a flat belt BDLR from a round or v-belt-driven unit is not recommended because the round or v-belt is prone to stretching and slipping and will not transmit enough power to reliably drive the larger BDLR.

6.1.1.2 Lineshaft Conveyor

Lineshaft conveyors are unique in that they offer the widest variety of bed types of any of the live rollers that we discuss. The carrier rollers are each driven by individual round urethane bands driven by molded plastic (acetyl) spools that fit onto a steel shaft running along the length of the bed (see Figure 6.6). The spools have a slip fit on the shaft and the tension from the urethane bands provides enough friction between the spool and shaft to drive the rollers. To increase driving force for heavier products, a shorter band can be used. For instance, if the standard band length is 267 mm (10.5 in.), a 260-mm (10.25-in.) band can be used to increase the friction between the spool and the shaft. Generally, the maximum driving force will move products that weigh 20 lb./ft. When more driving force is required, keyed spools are available. These can be used only if no accumulation will occur in the keyed area. If product accumulates in an area of keyed spools, one of two things will happen: The urethane bands will get stretched out and lose their driving capability or the bands will wear through the spools.

Lineshaft conveyors have various sections available such as straights (of course), curves, merges, junctions, and powered gates.

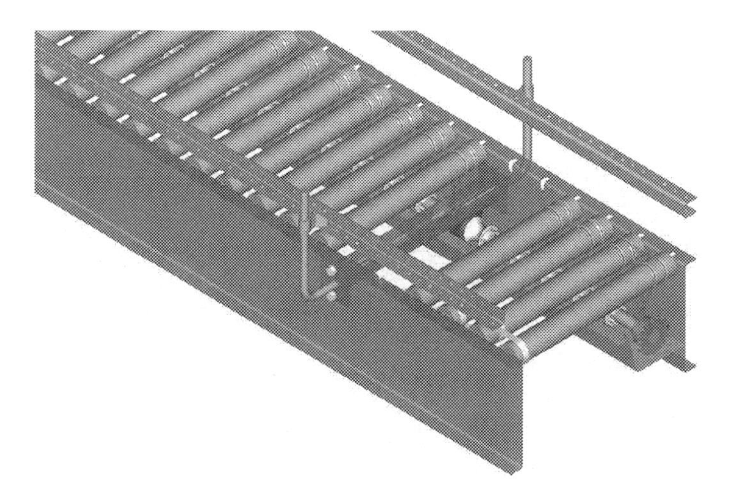

FIGURE 6.6 Lineshaft conveyor.

The straights come in lengths that are some increment of the rollers' centers, which are typically 3-, 4-, or 6-in. As with other live roller conveyors, far and away the most popular roller size is a 1.9-in. diameter. Some manufacturers offer a lineshaft conveyor with 1.4-in. diameter rollers that can be on 50-mm (2-in.) centers. This is very useful in handling smaller products. The 1.4-in. rollers are limited as to their length, typically to about 22 in. because the weight of the product will cause the roller to deflect too much.

Beneath each roller is a molded plastic spool that is connected to the roller by a urethane band. The band is typically a 4.7 mm ($^{3}/_{16}$-in.) diameter round urethane made into an endless band, something like a heavy-duty rubber band. A 6.3-mm (¼-in.) diameter band is available on heavy unit load applications using 55-mm ($2^{3}/_{16}$-in.) diameter rollers, but it is not very popular due to the limited drive capacity of the spools slipping on the lineshaft.

As the shaft turns, it is the natural tendency for the spools to move along the shaft and thus put undue strain on the bands. To prevent this movement, manufacturers use a variety of methods that keep the spools more or less stationary along the shaft. Some use snap-on plastic C-clips and others use thin-walled plastic tubing as spacers between the spools. Both work well, but each has a drawback. The C-clips must be moved to the other side of the spool if the conveyor is to be run in the other direction or clips must be placed on both sides of the spools if the conveyor will be bidirectional. The only real drawback of the spacer tubing is the additional time required to disassemble and reassemble the lineshaft for maintenance. Keeping all of the various length tubes in the right order can be tedious and time consuming. Thankfully, that only comes into play when maintenance requires that the shaft be removed.

The drive of a lineshaft conveyor can be anywhere along the length of the conveyor. It is never on a curve because the shaft on a curve cannot be under lateral loading. This will become apparent when the curves are discussed later. The drive is typically hung under the conveyor and consists of a motor and reducer simply driving the shaft, typically using roller chain and sprockets. Due to the simplicity of the drive design, many manufacturers offer a variety of drive configurations to match a specific application. Because no belt is being driven, there is no need for take-ups or belt tensioning.

Lineshaft conveyors have some unique advantages over other live roller conveyor types. One is the ability to drive rollers in both directions on the same drive at the same time. Simply by twisting the urethane band in the opposite direction around the roller, they will run in the other direction. This can be a real money saver in certain applications. A second unique feature is the use of speed-up spools. Some manufacturers offer larger spools to replace standard spools. This gives a lineshaft the ability to run at two different speeds on the same drive, which comes in handy when trying to create gaps between product.

6.1.1.3 Other Drive Methods

There are two types of padded chain drive live roller conveyors. As shown in Figure 6.7, one uses a roller chain that has hard plastic pads snapped into the chain between the links. The other uses a long, soft, perforated, urethane hose that sits over the extended pins of the chain. In both cases, the chain is pressed against the bottom of the rollers through the use of a wearstrip or channel guide rather than a series of sprockets.

Due to the chain being on its side, the urethane hose version allows straight and curved conveyors to be all driven by one chain.

Some manufacturers offer a light-duty chain-driven live roller (CDLR). CDLRs are discussed along with the heavier-duty versions in Chapter 7, "Heavy Unit Load Handling Conveyors."

6.1.1.4 Application Details

Live roller transportation is exactly what the name implies: The product is transported with little or no accumulation capabilities. Accumulation conveyors are discussed in Section 6.1.2.

Because of live roller conveyors' relatively low friction between the roller and the product being conveyed, they are preferred over belt conveyors where the product must be stopped momentarily, loaded or unloaded from the side, merged, diverted, or turn corners.

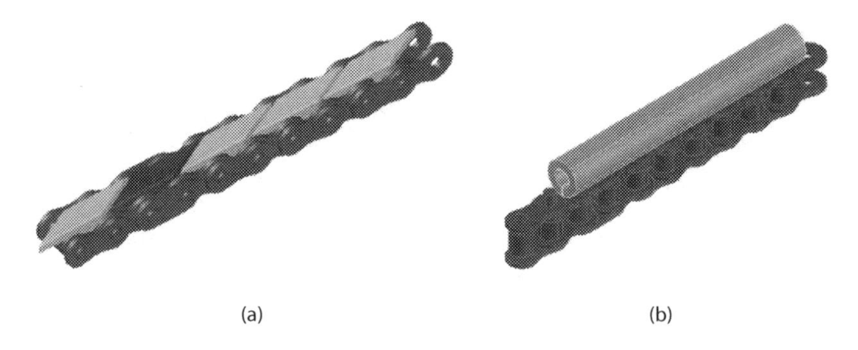

FIGURE 6.7 Padded chain.

BDLRs are not recommended for oily environments because the belt will lose its driving force on the rollers.

When multiple conveyor lines are brought together, such as prior to a sorting system, they must be merged together. Rather than requiring each conveyor line to have an angled merge on the end, many manufacturers offer what is known as a saw-toothed merge. This is typically a BDLR conveyor with one straight side frame and a series of steps in the side frame on the other side. This creates a series of entrances at a 45-degree angle from the direction of the take-away (see Figure 6.8).

When determining the width of the conveyor to be used, it is important to remember that not all manufacturers publish their conveyor widths the same way. For instance, one manufacturer says its conveyor is 450-mm (18-in.) wide, but actually it is 464-mm (18¼-in.) wide with a 406-mm (16-in.) wide roller surface. Watch out for nominal conveyor widths.

A conveyor should be wide enough to properly support all of the product to be carried. If the product is firm and flat on the bottom, then it is acceptable for the product to be wider than the rollers' surface. For instance, a carton that is 483-mm (19-in.) wide can be carried on the 450-mm (18-in.) conveyor just described. However, when the largest carton overhangs the roller surface, the narrowest carton cannot get more than halfway off of the rollers. In the preceding example, the widest carton is 483 mm (19-in.), add 25 mm (1 in.) for minimal clearance, and that leaves a 50-mm (2-in.) gap on either side of the roller surface: 483 + 25 – 406 = 102 mm

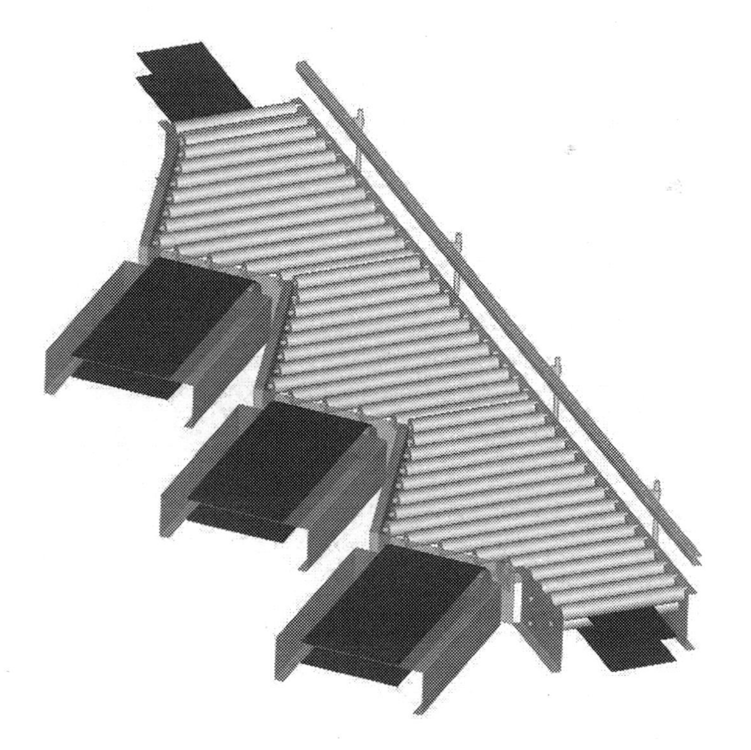

FIGURE 6.8 Sawtooth merge.

(19 in. + 1 in. – 16 in. = 4 in.); 102 ÷ 2 = 51 mm (4 in. ÷ 2 = 2 in.). If the narrowest product is less than say 125 mm (5-in.) wide, then a wider conveyor is required. It does seem odd, but the narrowest product can dictate a wider conveyor.

Beyond the obvious cost implications, there are other reasons to not just make all of your conveyors, for instance, 750 mm (30-in.) wide. When curves are introduced into a system, the package length can become a concern. Normally, the roller centers should be no more than one-third the length of the shortest product. A product that is 229-mm (9-in.) long needs minimally 75-mm (3-in.) roller centers. This rule ensures that there are at least three rollers under the product at all times. The challenge comes in the curves. Because the rollers of a curve are tapered and are arranged in a fan-shaped pattern, they are closer together along the inside rail (e.g., 64-mm [2½-in.] centers) and farther apart along the outside rail (e.g., as much as 150-mm [6-in.] centers). The roller spacing all depends on the curve width and radius. The larger the radius or narrower the width, the closer the rollers are along the outside.

As with any curve application, it is important to make sure that all the products that are to be carried on a conveyor system can navigate the curves properly. The following is a formula to determine the minimum conveyor width (see Figure 6.9). Frequently, due to longer products, the width or radius of the curve may have to grow. The formula is based on the Pythagorean theorem.

$$BF = \sqrt{(Radius + PackageWidth)^2 + \frac{PackageLength^2}{2}} - Radius + 1$$

The best way to approach the determining of conveyor width is to determine the minimum width of the conveyor based on the largest product and standard radius

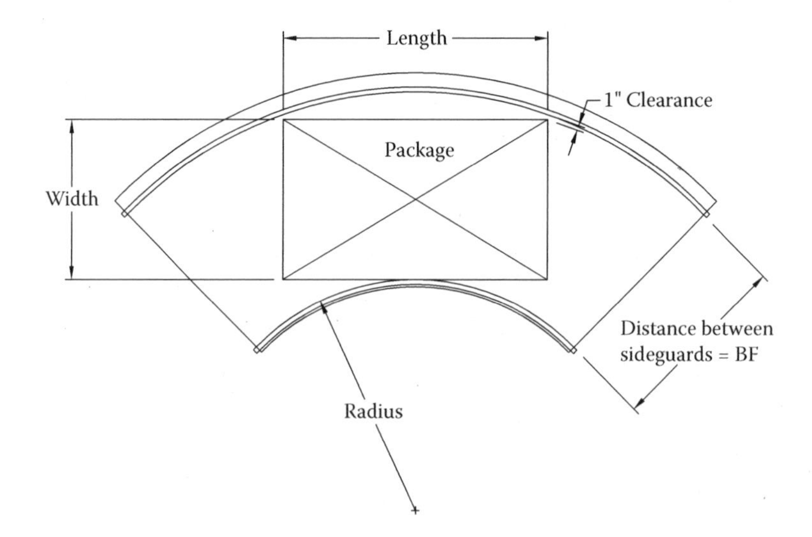

FIGURE 6.9 Curve sizing based on product size.

curves and then verify that the smallest product will move properly along the outside of the curve.

Careful application of the conveyor can ensure that the smallest product does not get to the outside of the curve except in rare occasions. Depending on the conveyor being applied, the rollers can be skewed slightly prior to the curve. This is done by shifting one end of the rollers back by 1½ in. to 3 in. so that the product is forced to one side of the conveyor. This can be done to keep smaller product toward the inside of the curves. If you have an S-curve type application, this, of course, will not work due to skewed rollers' inability to move a package from one to the other in a relatively short distance.

In many roller conveyors, the rollers can be skewed as shown in Figure 6.10. This forces all of the packages along one side of the conveyor. Some live roller drive types require modifications if rollers are to be skewed. The belt in BDLR will get off track due to the skewed rollers. Some manufacturers offer side guides for the belts in areas that have skewed rollers to keep the belt properly tracked.

Using L for the length of the skewed roller area, BF for between frames of the conveyor, PL for the average package length, RS for the roller skew, and TD for the travel distance from where the package is to the side,

$$\frac{RS}{BF} = \frac{TD}{(L - PL)}$$

Solving for the length gives us:

$$L = \frac{BF \times TD}{RS} + PL$$

There is one thing that is frequently overlooked when skewing rollers. The gap created between the last straight roller and the first skewed roller should not be more than 1/3 of the length of the shortest product being conveyed. If it is larger than 1/3, then the skew should be reduced or something must be added in the gap to support the product. Frequently, several skatewheels on a long ¼″ bolt are sufficient to fill the gap.

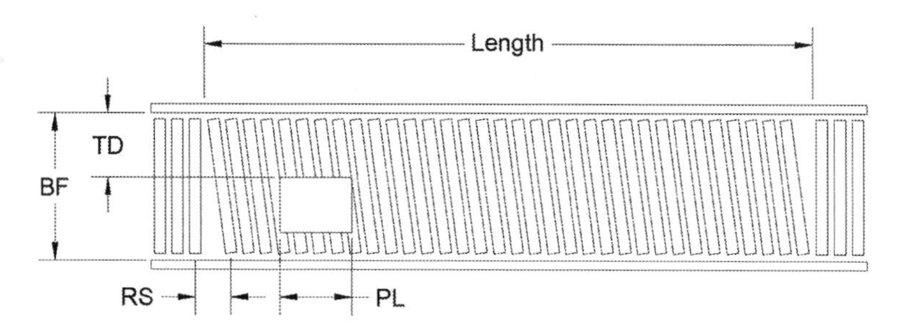

FIGURE 6.10 Skewed rollers.

6.1.1.5 Accessories

The following are among the accessories that are available for live roller transportation conveyors:

- Traffic cop: A traffic cop is a mechanical device used where two parallel conveyor lines must release product into a merge area without jamming. The traffic cop only allows product from one line or the other to be released at any given time. It works on a first-come, first-served basis.
- Pop-up transfers: These provide a method of moving products on and off the main line conveyor at 90 degrees. Transfers are typically available with either three to six urethane transfer belts, which are positioned to pop-up between the rollers. The urethane belts run in plastic tracks and on plastic sheaves and are either driven by a separate motor or slave driven when used in a lineshaft conveyor. For heavier products, the urethane belts can be replaced with roller chain.
- Product stops: Among their suitable applications, product stops help control merging or induction and can be used to initiate accumulation. The stop can be either a blade that pops up between rollers or replaces a roller with one that is built into the stop. Stops can be air actuated, motor driven, or manually actuated. Stops can be used to square products as well.
- Pop-up wheel diverters: A wheel diverter is factory mounted in a bed and typically has two rows of powered wheels positioned at an angle to the conveyor flow. With a pneumatic actuator, the wheels pop-up between the rollers to divert product onto an angled junction conveyor. These are discussed further in Section 6.3 later in this chapter.
- Roller brakes: Air-operated brakes push against the bottom of the rollers to make them stop turning. It is important to make sure that the rollers that will be pushed on by the brake cannot pop-out of the frame. Two brakes can be used in an alternating pattern to create an escapement for product spacing.
- Internal drive crossovers: These are used in lineshaft and round or v-belt-driven live roller conveyors to change the side of the conveyor that the driving force is on.
- External drive crossovers: These are used strictly on lineshaft conveyors to transmit power to an adjoining parallel conveyor.
- Gates: Gates are hinged conveyor sections used to allow personnel access to the other side of a conveyor (see Figure 6.11). Gates are typically counterbalanced using either heavy-duty springs or gas cylinders to make them easier to lift. Lineshaft conveyors and motorized roller conveyors can have powered rollers in the gate. All others typically use a gravity wheel or roller gate. For obvious reasons, gates cannot be equipped with pop-out rollers.

6.1.2 Live Roller Accumulation

Like the transportation versions, live roller accumulation uses the same rollers and many other similar components. The primary difference is that the accumulation conveyors have a method of stopping the rollers in a controlled fashion. For some types,

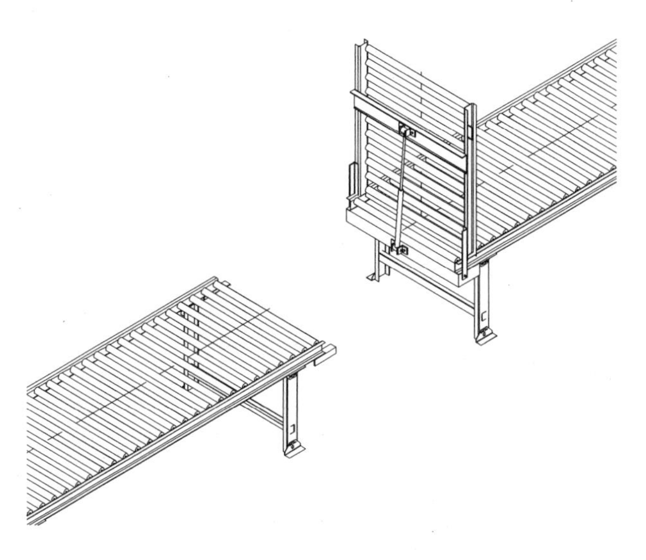

FIGURE 6.11 Roller gate.

it is a matter of using a brake to stop the rollers from turning, whereas others simply remove the drive pressure and still others actually control the drive in each zone. This is referred to as zero-pressure accumulation.

When discussing accumulation conveyors, the zone length is frequently considered. The zone is a group of rollers that are controlled together, stopped and started as a group. The zone is typically 24-, 30-, or 36-in. long, but, depending on the type of conveyor used, there are many other zone lengths available.

Regardless of the type of conveyor used, all have one thing in common: a method to sense the presence of product in a zone. This is done in a variety of ways.

The simplest but most expensive detection method is the use of a photo-eye. This electrical device is pointed at a target reflector. When something blocks the reflector, it sends a signal to whatever is controlling it. The photo-eye must be mounted in a position that allows it to see any of the products to be conveyed anywhere across the width of the conveyor.

A less expensive method of sensing product is through the use of a sensor roller. This is typically a smaller diameter, plastic roller that sits slightly above the rest of the rollers. When a product is conveyed over the sensor roller, it is depressed and it triggers a limit switch. The limit switch can be an electrical device, a small pneumatic valve, or mechanical linkage. The challenge is that all of the product to be conveyed must be heavy enough to depress the sensor roller. In some cases, we have seen very light product run up onto the sensor roller and stall, thus getting damaged when the next heavy product comes along. Another drawback to sensor rollers occurs when plastic totes or tubs are being conveyed. Due to the flange around the top of a tote or tub, it might run into the product in the zone ahead of it before it reaches the sensor roller. If this happens, when the next product comes along, it will not stop pushing on the tote, thus creating unwanted line pressure. If there are two totes, one can jump up on the lip of the other or the two can begin to bridge up from the line pressure. This, of course, can cause the product in the totes to be spilled and possibly damaged. When dealing with totes or tubs, many manufacturers use dual sensors, two sensor rollers linked together. Beyond the added cost, the other drawback to dual sensors is that the minimum product weight to trigger the sensor rollers is increased, typically doubled.

In recent years, another method that combines both of the previous two has been developed. Several manufacturers have developed a photo-eye that has an integral pneumatic valve. This eliminates the need for excessive wiring and is not dependent on the weight of the product. The photo-eye and valve combination has to be mounted in a position that allows the photo-eye to see any of the products to be conveyed anywhere across the width of the conveyor.

When dealing with a photo-eye for controlling accumulation, remember that the angle of the photo-eye to the conveyor is very important. If the photo-eye points across the conveyor, perpendicular to the product flow, it will more easily see any gap between products and actuate the accumulation. If the photo-eye is angled away from perpendicular, the trailing end of one product might not unblock the photo-eye before the next one blocks it. This will lead the system to think it is handling a long product and the accumulation will not activate properly.

We do not discuss low-pressure accumulation here because it really is simply a standard transportation live roller with reduced drive pressure on the rollers. This type of accumulation is acceptable for short runs and is not recommended for product that can be crushed or damaged, totes, or other taper-shaped products.

6.1.2.1 BDLRs

For the most part, BDLR accumulation conveyors utilize the same drives as the transportation version. The primary difference is that the pressure rollers are not mounted fixed in the frame but rather in a movable subframe that allows them to be dropped away, thus removing the drive from the rollers.

Most accumulation conveyors operate through the use of pneumatic devices to keep the driving belt pushing on the bottom of the rollers. A solenoid valve is typically used to control the first zone in a conveyor. When the valve is triggered, the drive belt drops away, removing drive from the rollers. The next product then coasts into the zone and triggers the sensor (roller or photo-eye). This in turn removes the drive from the preceding zone. The next product coasts into that zone and triggers the sensor, and the scenario is repeated again and again until all of the zones are full.

One manufacturer uses a 4-in. v-belt-backed belt to drive the rollers and a mechanical linkage to hold the belt up to the rollers. When a sensor roller is depressed, it kills the drive in that zone. The belt is dropped away just enough to remove the drive. In some applications, where it is desirable to have the product pushed close together, a short section of thicker belt can be added so that as it comes around under the rollers and it is just thicker enough to drive the rollers in the deactivated zones, thus pushing the product a few inches forward to close up any gaps.

Very few manufacturers offer a v-belt BDLR with accumulation, but several offer narrow-belt BDLRs. Instead of using rollers that support the belt, they typically use a shoe made of a low-friction material to hold the belt against the bottom of the rollers.

6.1.2.2 Padded Chain-Driven Live Roller

A padded chain-driven live roller conveyor is very similar to BDLR but instead of using a belt to drive the rollers, a roller chain with urethane pads snapped into it is used. The chain is a pre-straightened chain, so it does not curl or twist as it runs the length of the conveyor. Only one or two manufacturers produce this type of conveyor. It is included here because one of the manufacturers was, at one time, the world's largest manufacturer of conveyor systems, and there are hundreds of miles of this type of conveyor installed.

6.1.2.3 Lineshaft

The unique feature of lineshafts allows accumulation zones to be added virtually anywhere along the conveyor. Some manufacturers even offer accumulation through the curves. Accumulating in the curves does come with some risk of the product getting skewed during the start–stop action of accumulation, more so than on a straight conveyor bed.

The way lineshaft accumulation conveyors operate is slightly different from the BDLR type. Instead of removing the drive from a zone, the lineshaft uses an air-operated roller brake to stop the rollers from moving. There are several common brake designs that all work equally well. Stopping the rollers forces the spools to slip on the lineshaft. A solenoid valve is typically used to arm the first zone in a conveyor. The next product triggers the sensor (roller or photo-eye), the brake is engaged, the rollers are stopped, and the preceding zone is armed. The next product triggers the sensor, the brake is engaged to stop the rollers, and the scenario is repeated again and again until all of the zones are full.

6.1.2.4 Motorized Roller

The final type of live roller accumulation is the motorized roller. This type of conveyor is driven by small motors built into some of the rollers (Figure 6.12). It is frequently referred to as MDR conveyor, short for motor driven roller conveyor. Using

FIGURE 6.12 Motorized roller.

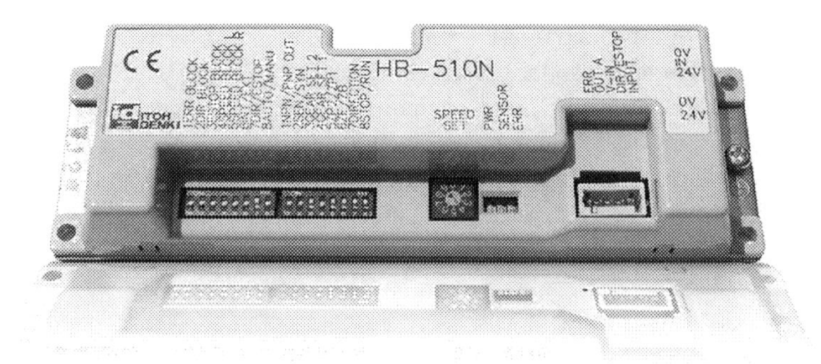

FIGURE 6.13 Motorized roller controller.

urethane bands similar to those used in lineshaft conveyors, one motorized roller will slave drive a series of several more rollers to create an accumulation zone. Each zone will have a dedicated photo-eye to sense product on the conveyor. The rollers are available in 24V DC, 48V DC, and 120V AC. The 24V DC rollers offer better speed control and higher speeds and carrying capacity than the 120V AC rollers.

One of the advantages of motorized roller conveyors is that it is easy to create a conveyor that is only 24″ long without an oversized motor and reducer.

Each motorized roller has a controller that it is plugged into. Some controllers are designed for transportation only and others for accumulation. There are some controllers that offer individual programmability or control more than one roller (Figure 6.13).

Most controllers have speed settings as well as a series of small dip switches that set various parameters to dictate how the controller responds to certain conditions.

MDR conveyors offer some of the quietest operation due to the individual zones shutting off when there is no product flowing. This also reduces noise and energy consumption significantly.

Back in Chapter 4, we mentioned MDR belt conveyors. They are MDR roller conveyors where belts have been stretched over all the rollers in a zone. These belted zones are great for handling small or flexible products such as polybag mailers. They can also be used as metering belts through an operation such as applying labels to a box.

A close cousin to MDR conveyor is the low-voltage conveyor made by Hytrol that uses a small 24V pancake motor mounted under the rollers to drive them like a motorized roller.

CASE STUDY – PHOTOGRAPHIC PAPER ROLL HANDLING

MDR offers great flexibility and can help solve some very unique problems. Some time ago while working for Rapistan, I was tasked with developing an accumulating roller conveyor with some very stringent requirements. The product being handled was rolls of photographic paper. The conveyor system was to be installed in a dark room. This is not like the dark rooms on television where there is a red glow all around but so dark you literally cannot see your hand in front of your face.

First, the rollers had to be stainless steel. There could be no rust allowed because it may transfer onto the photographic paper. The chemicals in galvanizing and zinc plating would affect the paper.

Second, the conveyor had to be intrinsically safe. Meaning anyone could put their hand on it anywhere without getting hurt.

Third, the rollers had to be very straight. All rollers are slightly banana shaped. No manufactured tubing is perfectly straight. The straightness tolerance was a total indicator runout of 25.4 µm (0.001"). If the roll of paper bumped into a roller that was bending upward, it could bruise the paper and cause the pictures printed on it to be cloudy.

Lastly, the roller surface could not get hot because it would damage the paper. This would cause the pictures printed on the paper to be cloudy.

The intrinsically safe requirement led us to either lineshaft or motorized roller conveyor. Lineshaft with the spools' ability to slip offered the safety feature required, as did the low drive force of the motorized roller. There were no pinch points with either design as well as long as we stayed with round urethane bands. Motorized rollers are available with poly V-belts. These are very tight and can cause serious pinch points.

The tight straightness tolerance eliminated the lineshaft conveyor. The drawback of the lineshaft conveyor is that the grooves for the bands are rolled into the shell of the roller. This process introduced more inconsistent straightness in the rollers. A standard offering for motorized rollers is to have machined end caps with the grooves in them, rather than the grooves rolled in the shell. Figure 6.14 shows two motorized rollers, one with the grooves in the shell and one with the grooves machined into the end cap.

As no one made such a conveyor, we had to figure out how to manufacture the conveyors ourselves. We discussed these requirements with the two major motorized roller manufacturers at that time. One could not accommodate the straightness requirement. The other, Itoh Denki, said they could hold the required tolerances and provide stainless steel shells and machined endcaps. Itoh provided all of the motorized rollers as well as the unpowered idlers; so, everything matched.

We decided to use standard conveyor frames from Rapistan's model 1276 padded chain-driven live roller that we could order from the factory. This also provided us with integral side guides (Figure 6.15). We mounted the motorized

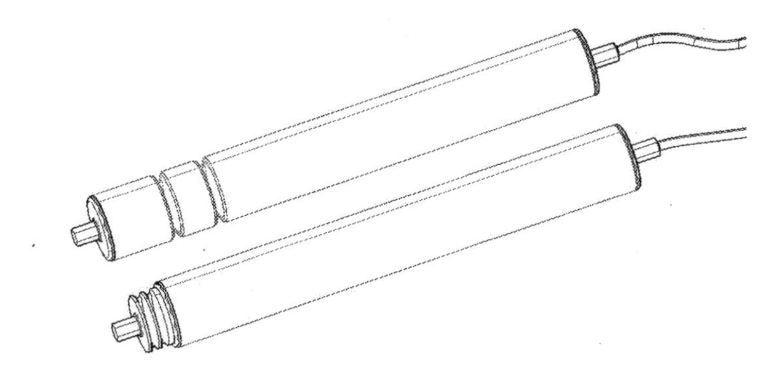

FIGURE 6.14 Grooved rollers.

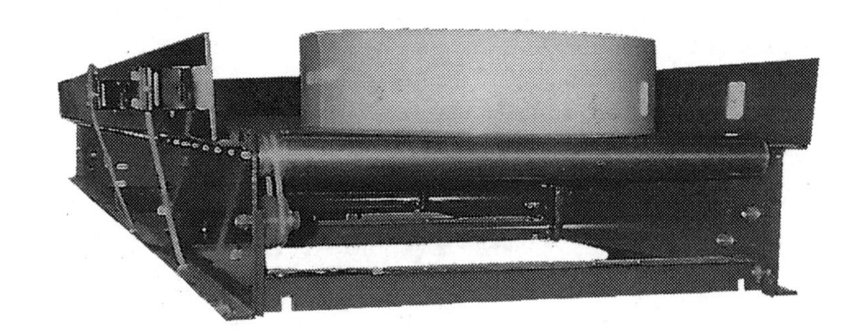

FIGURE 6.15 Conveyor.

rollers and added all the wiring to create the accumulation conveyor with 24″ zones. Each zone consisted of one motorized roller and seven idler rollers slave driven from the motorized one. To further protect the product, we added 1 ½″ (38 mm) open cell foam behind ultra-high molecular weight polyethylene (UHMW) on all stops so that the product didn't bump anything to bruise the paper.

The final challenge was to control the coasting of the product after the rollers were shut off. The motorized rollers can be equipped with a brake, but that generated too much heat in the rollers. We then developed a spring-loaded drag wheel that was mounted beneath the last roller in each zone. This allowed us to apply sufficient drag to stop the rollers faster without generating too much heat in the drive roller.

6.1.2.5 Accessories

Live roller accumulation conveyors utilize the same accessories as their transportation counterparts. The exception is that you'll never see a pop-up wheel divert in an accumulation conveyor.

6.1.2.6 Application Details

Accumulation conveyors are used when feeding into a machine or a sortation system. Because of how basic accumulation logic works, many manufacturers have developed modified versions for specific applications.

Because the accumulating action of the rollers kills the drive to the rollers in the previous zone, the products end up with gaps between them equal to at least the zone length. This results in a throughput of half of what would be if the product was back-to-back. Because the product is moving along in a separated manner, this is referred to as singulated. This gapping feature allows photo-eyes to see between products to ensure that product is moving smoothly and that there are no jams. If there were no gaps, three products tight together might block a photo-eye on the infeed end of a conveyor long enough to make the photo-eye think the conveyor was full and shut off the upstream conveyors or machine, when, in actuality, the conveyor might have been mostly empty. This reduced throughput forced the development of two additional types of accumulation logic: dynamic and slug (see Figure 6.16).

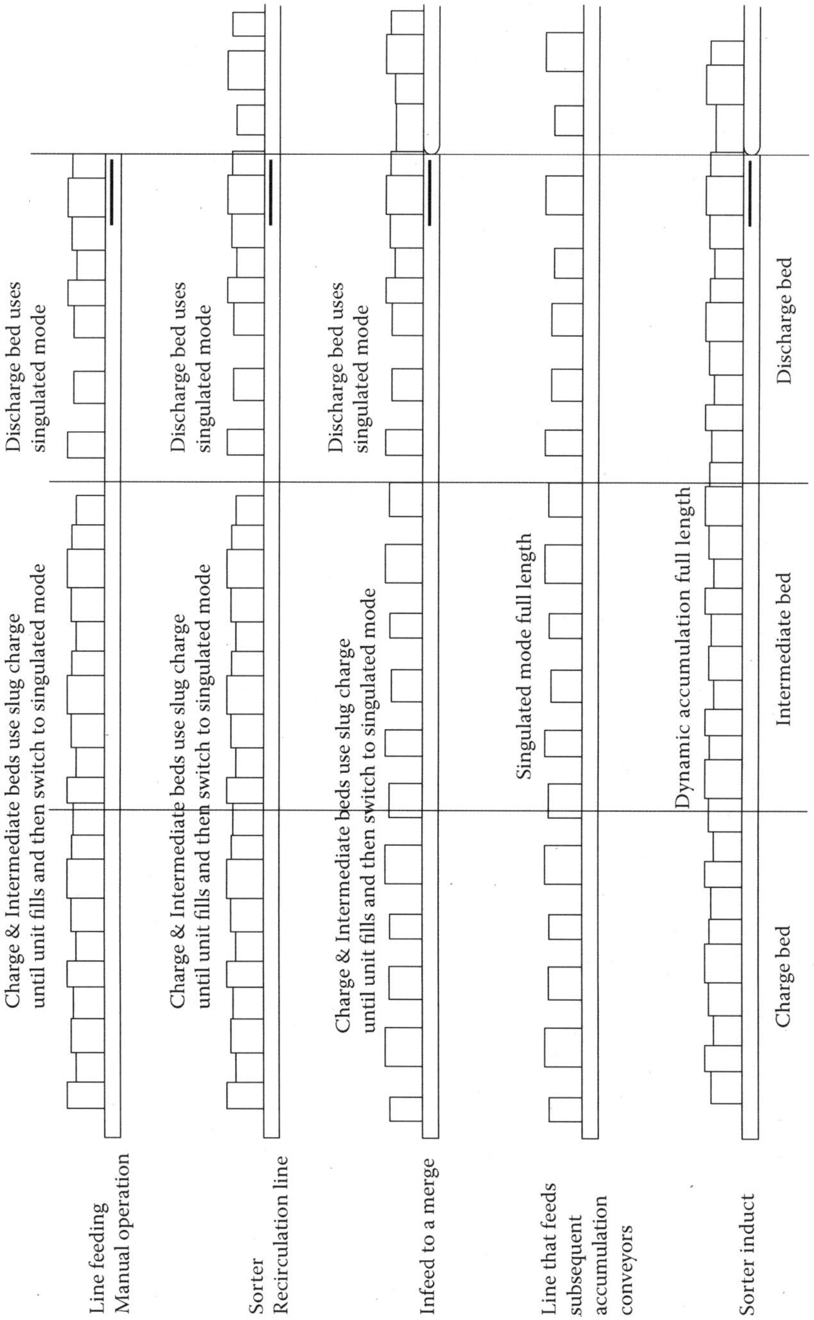

Line feeding
Manual operation

Charge & Intermediate beds use slug charge until unit fills and then switch to singulated mode

Discharge bed uses singulated mode

Sorter
Recirculation line

Charge & Intermediate beds use slug charge until unit fills and then switch to singulated mode

Discharge bed uses singulated mode

Infeed to a merge

Charge & Intermediate beds use slug charge until unit fills and then switch to singulated mode

Discharge bed uses singulated mode

Line that feeds subsequent accumulation conveyors

Singulated mode full length

Sorter induct

Dynamic accumulation full length

Charge bed

Intermediate bed

Discharge bed

FIGURE 6.16 Accumulation types.

Dynamic accumulation reduces the drive pressure when accumulating rather than removing it completely. This turns the accumulated zones into a low-pressure live roller, thus keeping a very low line pressure on the product to ensure that there are no gaps. This is frequently applied on the last accumulating conveyor prior to a sorter induct.

Slug accumulation can be applied at the infeed or exit end of a conveyor or the entire length. The slug feature basically disarms the accumulation capability and temporarily turns the conveyor into purely transportation.

This can be used on the infeed end of an accumulation conveyor so that the product does not slow down or singulate. This is important at the discharge of a process machine or high-speed sorter.

Slug accumulation can be used on the exit end to allow product to discharge without gaps as a long slug. This is frequently used where the rate entering the conveyor is low and the exit rate is high, such as feeding into a saw-toothed merge. All of these different types of accumulation are controlled by solenoid valves.

6.2 MODULAR CONVEYOR

Modular conveyors are a unique group of conveyors that do not really fit into any other general group. They are most frequently used for in-process manufacturing systems and utilize a high level of control hardware. The conveyor itself typically has an extruded aluminum frame with wearstrips to guide a chain or belt. This extrusion also aids in the mounting of accessories. Some also have built-in wireways for the control wiring and pneumatic headers to feed the air-operated devices.

6.2.1 Multistrand Belt Conveyor

Multistrand belt conveyors are typically two strands of narrow flat belts, frequently long-timing belts. Due to their narrow width and relatively long length, these belts typically use Kevlar or steel cable as a reinforcing member.

6.2.2 Multistrand Chain Conveyor

Multistrand chain conveyors typically consist of two strands of chain to be used as the conveying surface. The chain can be in a variety of forms: standard roller chain, roller chain with plastic pads snapped into it, or chain with oversized rollers. Figure 6.15 shows five of the commonly used types of chains. The first is a padded chain, a standard chain with low-friction plastic pads snapped into the space between the rollers of the chain. The pads are in every other space so that the sprocket teeth can fit into the open spaces. When the pallets accumulate, the low-friction pads allow the chain to slide under the pallet. The only drawback, and this holds true for all of these chains, is that if the pallets do not have a chamfered or radiused edge, the pallets will tend to vibrate. This can be a problem for the product on the pallet if it is not contained. The next three types carry the pallets on the rollers; the chain is supported by the small center rollers and is pulled in the direction of the pallet flow. The product moves at the speed of the chain, and the rollers allow the product to accumulate while

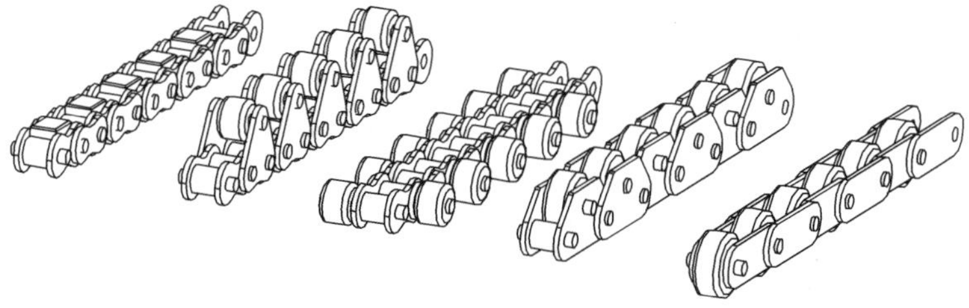

FIGURE 6.17 Various conveyor chains.

the chain continues to run underneath. The rollers minimize the line pressure. The final chain operates in a completely different manner. The chain moves in one direction riding on the large rollers, and the pallets ride on top of the rollers moving in the opposite direction at twice the chain speed. The speed differential is created by the size of the rollers. Systems that use both belts and chains need to make accommodations for the difference in their thicknesses (Figure 6.17).

6.2.3 Curves

To change direction in a modular system, three primary methods are used. The first is a live roller curve with tapered rollers. These are typically specially designed with a very tight inside radius to minimize the space requirements. The roller curve is the least expensive of the three choices. The second is a turntable. The turntable offers a smaller footprint for making 90-degree turns, and there is no concern about the product getting out of orientation, as there can be with roller turns. The third and final method is a belt turn. Like the turntable, there is no concern for orientation and belt turns can be designed with a very small inside radius. The belt turn has the smallest footprint of the three but is frequently the most expensive alternative.

6.2.4 Accessories

Each manufacturer offers a variety of accessories for their own system. These include special parts bins and operator workstations that mount to the conveyor, as well as a plethora of other options. The following and their functions are the three most common.

- Pallet stops: These devices are used to stop a product on the conveyor at a workstation, at a transfer, or for accumulation purposes.
- Pop-up transfers: These devices are typically mounted between the strands of belt or chain and consist of two narrow belts. The product is stopped over the transfer and the transfer unit is raised, typically by an air-operated actuator, the belts are started, and the product is transferred off onto another conveyor or into a workstation.
- Pop-up turntables: These are used to lift a product off of the conveyor and rotate it either for an operation to be performed on it at the turntable or to be set back down in a new orientation in preparation for some downstream operation.

6.2.5 Application

On most modular conveyor systems, the product is carried on a slave pallet that is typically reused many times. As mentioned before, modular conveyor systems are used for in-process manufacturing systems. One of the reasons is that the system can be rearranged for a new system. When one product has run its life cycle, the conveyors can be reconfigured for a different product's manufacturing process. Due to their inherently tight tolerances, modular conveyors are frequently used for automated systems with robots performing some of the operations.

Modular conveyors can be used for mixed-model assembly operations. For instance, a company can assemble a series of different printers on a single conveyor system. Adding a radio frequency identification (RFID) tag to each pallet will indicate what type of printer the pallet has onboard, so the control system can route the pallet to the correct workstations. The advantage, in this case, is that many printers go through many of the same processes. Pallets with printers that require a unique process are diverted to that workstation and others that do not need that process pass by (see Figure 6.18).

Modular conveyor systems are more expensive than typical traditional conveyor systems because of the heavy controls content. The savings come in two forms: The installation is quicker and less expensive, and the modular conveyors offer flexibility and reusability of the equipment. With one system reconfiguration, the higher additional cost of the original system is small compared with the cost of a second new traditional system.

Some manufacturers have been trying to apply this concept of modularity to their full product line. This concept also has a higher initial cost for the equipment, but installation is much quicker and less expensive.

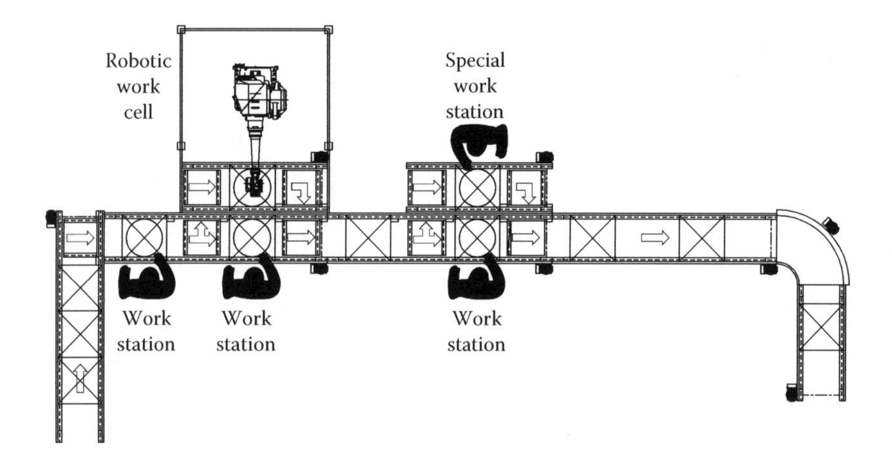

FIGURE 6.18 Example of a modular conveyor system.

6.3 SORTERS

Sorters accept a steady stream of product and divert it to its proper destination. Sorters are a unique type of conveyor in that although many are based on other conveyor types, many are unique. Therefore, we felt it necessary to handle them separately. Regardless of type, they are all applied in the same general manner. The differentiation comes in their rated capacity and the products that can be handled.

All sorters have two things in common. First is a method of feeding the product into the sorter in a properly oriented and spaced fashion. Second are receptacles for the product as it leaves the sorter. These can be gravity or powered conveyors or chutes.

The sorter inducts, those conveyors that feed the sorter, have one purpose: to feed product into the sorter with the proper orientation and spacing for the sorter to operate properly. The induct could be as simple as two roller brakes operating as an escapement on a lineshaft conveyor. It can also be as complicated as multiple servo-driven metering belts that feed a merge bed, which in turn feeds the sorter.

The sorter induct frequently includes a scanner or camera system to identify the product and determine its destination. Occasionally, scales and dimensioners are included as well as a quality check. If a box weighs too little, it may indicate that an item on an order wasn't in the box. That box can be diverted to a hospital or reject destination for inspection or correction.

6.3.1 Paddle Sorter

Paddle sorters use a pivoting diverter arm that is mounted beside the conveyor and forces the product off to the side (see Figure 6.17). These are good for lower-speed applications, and they are usually used for irregularly shaped units such as bundles or mailbags. Paddle arms themselves are typically actuated using an air cylinder or an electric motor. The paddle is actuated when the product is in front of the paddle. The cycle time is relatively slow; therefore, this type of sorter is typically used in only low-rate applications.

6.3.2 Pusher or Puller Sorter

Pusher or puller sorters use devices mounted to the side of a conveyor to divert cartons at a 90-degree angle with the use of an air cylinder (see Figure 6.18). Both belt and live roller conveyors are used for this type of sorter. The belt conveyors offer better product location tracking, but the side force of diverting the product can cause the belt to come off track. Live roller conveyors offer better side movement on the rollers, but some of the product tracking accuracy is lost due to roller slippage under the product. This type of sorter is good for large, heavy, and stable units. A pusher or puller is not a good choice for fragile units due to the somewhat rough nature of the diverting action.

A few manufacturers offer pusher or puller sorters for small, lightweight products such as boxes of checks or other small packages. These are typically built around a flat belt conveyor, but the same still holds true: They are not recommended for fragile products.

Keep in mind that because the product is being pushed off at a 90-degree angle, the conveyor receiving the product must be wide enough to handle the longest product and the rollers must be close enough for the width of the narrowest product to be carried in a stable manner. This type of sorter is a low to midrange-rate sorter with a rate of up to 40 product units per minute.

6.3.3 Pop-Up Transfer Sorter

Pop-up transfer sorters use belts or chains that are raised up to come in contact with a unit and divert it at a right angle (see Figure 6.19). Depending on the type of conveyor in which the transfer sorters are mounted, they can divert to one or both sides with the same unit. CDLR is the only conveyor that prevents two-sided sorting. In lineshaft conveyors, the transfers can be slave driven from the lineshaft. All other roller conveyors require the transfers to have their own drive. Orientation of the product ends up 90 degrees from the previous direction of travel. Transfer sorters are considered low-rate sorters, with a maximum recommended rate of 20 to 25 product units per minute.

There is also a variation on the pop-up transfer sorter. This variation does not pop-up but instead has a raised section on the transfer belts that acts as a pusher as they turn around the end pulleys to push the product off the conveyor.

6.3.4 Lineshaft Pop-Up Wheel Sorter

As mentioned previously, lineshaft conveyors can have a pop-up wheel diverter mounted in them. A series of these diverters can be applied to create a sorter (see Figure 6.20). Typically, the wheels of the diverter are slave driven by the lineshaft.

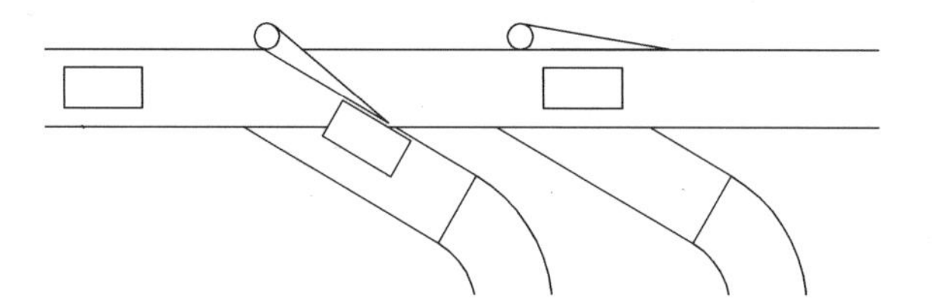

FIGURE 6.19 Paddle sorter.

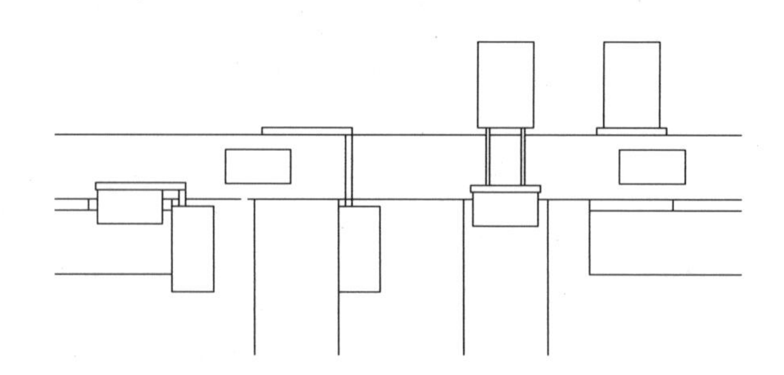

FIGURE 6.20 Pusher/puller sorter.

Lineshaft sorters can also power the sort spurs to take the product away as well. Lineshaft sorters are effective at rates below 50 product units per minute. Product should be aligned along the divert side of the conveyor.

6.3.5 High-Speed Pop-Up Wheel Sorter

High-speed pop-up wheel sorters are typically a series of pop-up wheel diverters incorporated in a flat belt conveyor (see Figure 6.21). The diverters are usually slave driven from the belt. Product needs to be aligned to the divert side of the conveyor. High-speed pop-up wheel sorters are considered midrange sorters with a rate of up to 125 product units per minute.

6.3.6 Pop-Up Steerable Roller Sorter

Like the high-speed pop-up wheel sorter, the steerable roller sorter is a series of slave-driven diverters incorporated into a flat belt conveyor (see Figure 6.22). The primary difference, as the name implies, is that as the rollers are popped up, they are turned to the left or right to provide sortation to both sides of a conveyor. Depending on the width of the conveyor and the products to be handled, there will be a justification diverter at the infeed end of the sorter. Its purpose is to move each product over to the side of the conveyor off of which it will ultimately be diverted. If the products are all large enough, they can be run centered down the conveyor and still divert properly. Pop-up steerable roller sorters are also considered midrange sorters with a rate of up to 125 product units per minute (Figures 6.23, 6.24).

6.3.7 Narrow Belt Sorter

There is a sorter that offers the continuous conveying surface of a belt conveyor with wheel or roller diverts (see Figure 6.25). The narrow belt sorter (NBS) uses six or

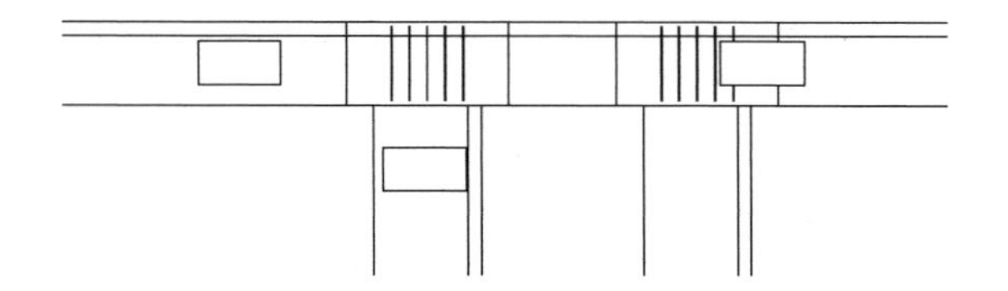

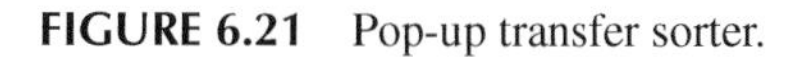

FIGURE 6.21 Pop-up transfer sorter.

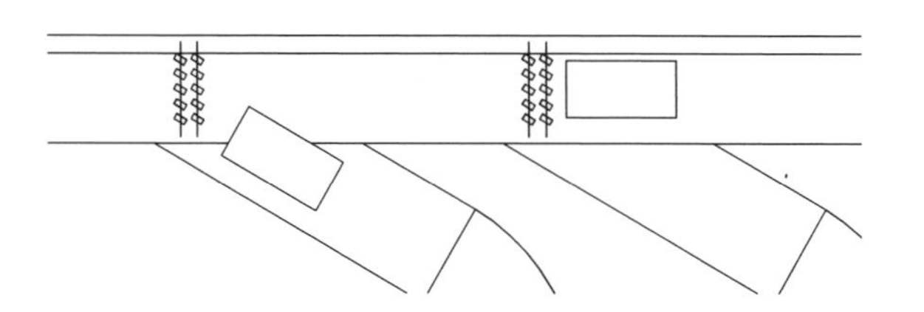

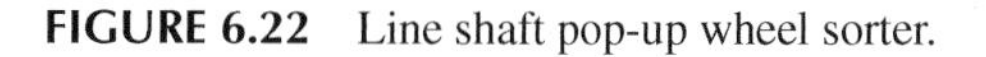

FIGURE 6.22 Line shaft pop-up wheel sorter.

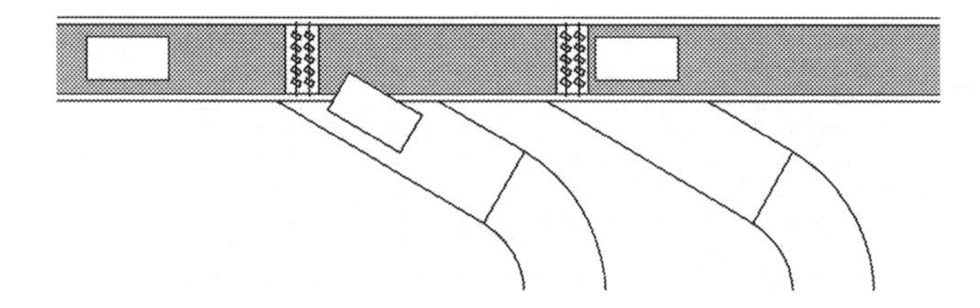

FIGURE 6.23 High-speed pop-up wheel sorter.

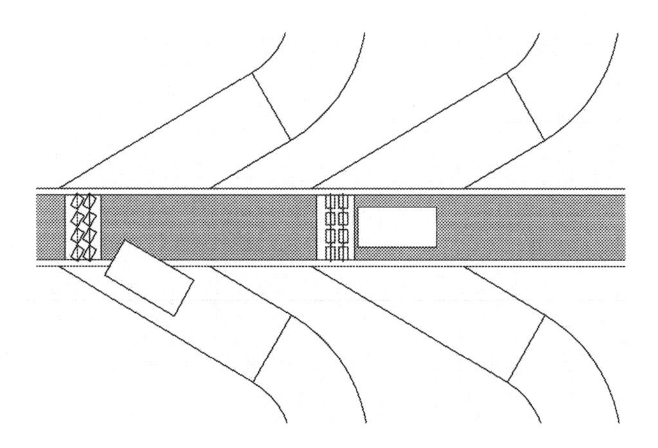

FIGURE 6.24 Pop-up steerable roller sorter.

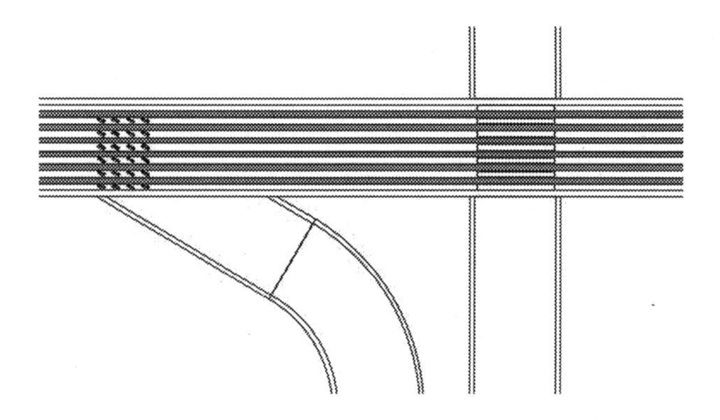

FIGURE 6.25 Narrow belt sorter.

more narrow belts for the conveying surface and can run at speeds up to 1.5 m/sec (300 FPM). The diverts are arranged so that the wheel or rollers pop-up between the belts.

The wheels are limited to diverting off one side or the other while the 90° roller can divert off of either side. The wheels and rollers have a high-friction surface to ensure that the product is diverted.

The wheel diverts are available in a variety of configurations depending on the manufacturer. The wheels are arranged on rows and there can be as few as three rows or as many as 12. One manufacturer offers a 12-row divert that operates in a wave fashion. The wheels are divided into three groups that operate sequentially like a wave. A typical divert has to keep all of the wheels raised until the product has successfully exited the conveyor. With a wave divert, the first group of wheels can drop as soon as the box is off of them while the others are still raised. Then the second one drops while the third remains raised until the product has successfully exited the conveyor. This allows the boxes to be closer together, thus increasing the rate through the sorter. The 30-degree wheel divert can attain rates of 100 cases per minute, and the wave version can achieve rates up to 200 cases per minute.

The 90-degree diverts pop-up to transfer the product off the side of the conveyor. They transfer the product on the fly, meaning the sorter doesn't pause once the product is centered on the divert. With these diverts the sorter can handle rates of 50–65 cases per minute. At least one manufacturer offers a smaller, narrower version for sorting small items. It can achieve rates of up to 100 items per minute.

6.3.8 Positive Sorters

Positive or sliding shoe sorters are different from all of the previous sorters we have discussed. The previous sorters have all been based on a standard conveyor acting as the host – be it flat belt conveyor or live roller transportation – that has devices added to give it the capability of sorting. Positive sorters are complete, self-contained units. Positive sorters are designed so individual sorter shoes traverse from one side of the conveyor to the other and in doing so push the product off the side onto a take-away conveyor or chute (see Figure 6.24). The conveying surface is made up of either tubes or, more commonly, extruded aluminum slats. The tubes or slats are attached to a heavy roller-type chain. The tubes, due to their spacing, are limited on the small end as to the product size they can convey. For product stability, the product must span at least three tubes, or be 9-in. long. Slats allow for a much broader product size and can sort virtually any conveyable mix of product. In general, sorters of this type offer superior rate and carton-handling capabilities. Positive sorters are considered high-speed sorters, with rates that range from 150 up to 300 product units per minute (Figure 6.26).

The drives for positive sorters are generally at the discharge end of the sorter and use two large sprockets to engage and drive the chains to which the slats are attached.

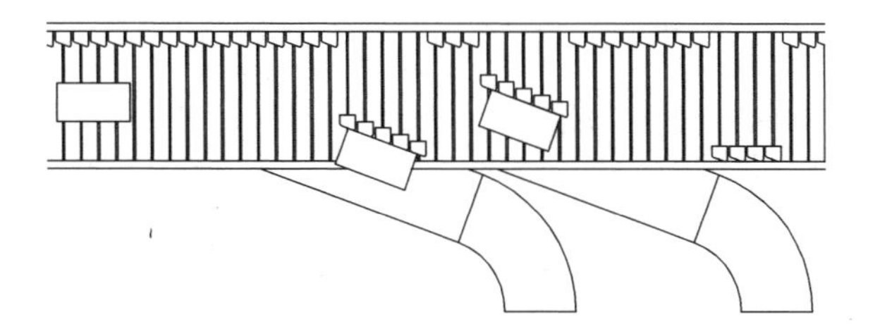

FIGURE 6.26 Positive or sliding shoe sorter.

Due to the cost of this type of sorter, they are generally used only in larger systems and therefore they are usually quite large. It is not out of the ordinary for positive sorters to have 5, 7.5, 10, or larger horsepower drives. The drive is typically its own module, rather than a bolt-on section of the conveyor. The end drive usually also houses any lubrication equipment to keep the chain and the shoes properly lubricated.

The diverts in a positive sorter are not actually visible in the system. Underneath the slats, there is a track in which a pin on the bottom of the shoe rides. At the divert positions, that track has a solenoid that diverts the pins into a diagonal track, thus sliding the shoe across the slat. Anything riding on top of the slat is then pushed along as well until the shoe reaches the other side and the product on top is pushed off of the conveyor.

Depending on the sorter design, the diverts can be designed to push to either side. This, of course, requires that the shoes be on the correct side of the conveyor when they come up at the tail end and the product enters the conveyor. On the under or return side of the sorter, typically just before the slat comes back up on top, there is a deflector that pushes all of the shoes to one side. This ensures that all shoes are in position on the side opposite the receiving conveyors or chutes. If the sorter is designed to sort off both sides, the fixed deflector is replaced by a divert that puts the shoes on each side as required. The number of shoes put on each side by this divert is determined by the sort controller and is based on the size of the product about to enter the conveyor.

One serious consideration when dealing with positive sorters is maintenance access. These machines require a high level of accuracy and tight tolerances to keep running. It is important to make sure that if your organization does not have the technical ability to maintain the sorter properly, then the manufacturer or local representative offers that service.

6.3.9 Recirculating Carrier Sorters

There are several types of recirculating carrier sorters. However, all have similar functions with carriers that handle the product to be diverted running in a continuous loop.

Recirculating carrier sorters can use manual or automatic inputs of product onto the trays, although most are fully automatic induction systems. Because of the continuous loop configuration, these sorters do not use a typical inline induction system to introduce product to the sorter. Typically, with this type of sorter the induction process involves two or more angled belt merges that convey the product onto the waiting tray or belt. Each induct point can process up to 60 product units per minute. The tilt tray and the cross-belt can both operate at speeds up to 5 m/sec (1,000 FPM). Both of these sorters are considered high-speed sorters, with rates that range up to 200 product units per minute for the tilt tray sorter and up to 300 product units per minute for the cross-belt sorter.

Some manufacturers offer a feature that is unique to recirculating carrier sorters. That is the ability to change elevations through inclined and declined sections of track. Be sure that the product cannot slide off the back of the carrier as it navigates the inclines and declines, keeping the angle of incline fairly shallow.

Recirculating carrier sorters are particularly adept at handling product that would be considered non-conveyable on other sorters. The following is a brief list of some of the more popular applications for this type of sorter.

- Letter trays, packets, and small envelopes in the postal and courier industry.
- Shoeboxes and apparel in the footwear and clothing industry.
- Plastic film-wrapped products with high-friction (sticky) surfaces, as in mail-order companies and the newspaper distribution industry.
- Unstable and sensitive products, such as frozen foods, cakes, and dairy products.
- Other industries with extreme variation in product types, shapes, and frictional behavior, such as baggage handling in airports or freight centers.

6.3.9.1 Tilt Tray Sorters

Tilt tray sorters (see Figure 6.25) are suitable for a wide range of products. The sorter can discharge product on either side. Larger products can occupy two trays, and they can be programmed to actuate in a sequence that will ensure proper orientation of large products. Because large products can be handled on two trays, the tray pitch, the center-to-center dimension of trays, can be minimized rather than requiring a tray to be big enough to handle the largest product. The sorter can discharge product into sloped chutes or directly into boxes or totes. Typically, these units are designed for lightweight, high-speed applications, but some units are designed to handle loads up to 75 kg (165 lb.) (Figure 6.27)

Tilt tray sorters are typically driven individually by linear motors mounted on each carrier, and subsequent carriers are all tied together to form a continuous loop. There are a variety of other drive concepts available, but none are as widely used.

It bears noting here that a few small manufacturers built a tilt tray sorter that has the trays and tilting mechanism mounted to a belt. In this case, the sorter is not a continuous loop but simply a straight sorter. Any product that does not get diverted

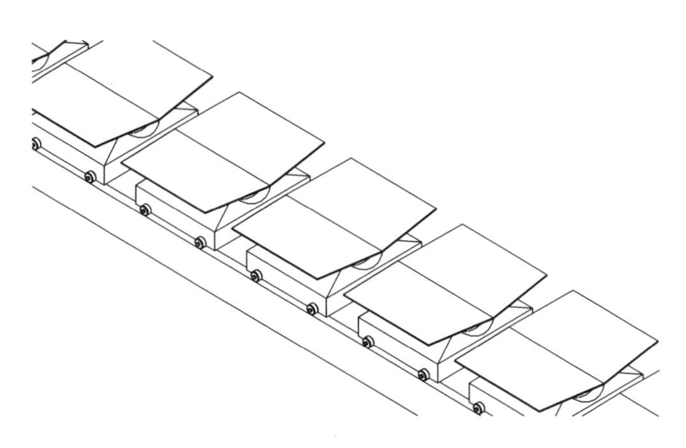

FIGURE 6.27 Tilt tray sorter.

must be re-inducted rather than being allowed to recirculate. At the end of the sorter, the trays follow the belt underneath through the return to the infeed end of the sorter.

Two aspects of a tilt tray sorter require attention. One is the material of which the tray is constructed. Some are made of plastic, and others are made from plywood. Plastic works well for small, lightweight product but can build up static when conveying plastic products. In this case, it is important that there is a grounding method. Plywood trays typically cost more and use high-grade plywood because of its superior multi-ply construction. These plywood trays are used for heavier applications and do not have the static issue of some plastic trays.

The second aspect that requires attention is the tilting device. There are many tilting devices on the market. Mechanical trip levers that push the tray up are recommended for only the simplest of applications. More popular is electric-operated actuation. This gets its power from the same source as the linear motors, so it is available anywhere along the system. The other reason this is more popular is that the tilting action is separate from the tray location and therefore can be optimized to increase the throughput of the sorter.

6.3.9.2 Cross-Belt Sorter

Cross-belt sorters are fed in the same manner as tilt tray sorters with multiple induction points. The primary difference is instead of trays that tilt, these sorters have a series of independently operated belt conveyors that are transversely mounted on a track (see Figure 6.28). To induct a product, the specific cross-belt is activated to provide a controlled transfer.

Cross-belt sorters are widely used in the retail trade, in food handling, and for e-commerce. These units are typically designed for lightweight, high-speed applications, but some units are designed to handle loads up to 50 kg (110 lb.).

Typically, cross-belt sorters are also driven individually by linear motors mounted on each carrier and subsequent carriers are all tied together to form a continuous loop. Again, there are a variety of other drive concepts available but none that are as widely used.

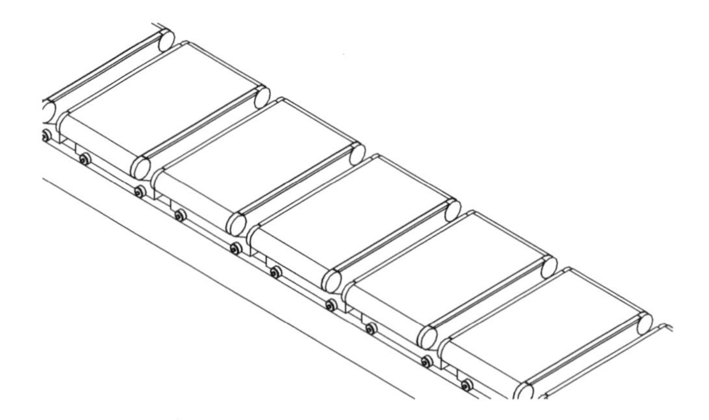

FIGURE 6.28 Cross-belt sorter.

Some manufacturers offer a double-deck sorter, which can double throughput in the same space or allow you to achieve the same throughput in half the space if space is at a premium.

6.3.9.3 Pusher Tray sorter

Another recirculating sorter that is used for soft, small, or irregularly shaped items is the Pusher Tray sorter. These are often used for garments in bags. The Pusher Tray sorter, as you would gather from the name, is a series of trays, each with its own pusher bar mounted across the width of the tray. When the tray has reached its designated discharge point, the pusher bar sweeps across the tray, pushing out the contents into a chute or tote (Figure 6.29).

6.3.9.4 Bomb-Bay sorter

Finally, we have the bomb-bay sorter. As the name infers, the carrier is a split tray and when it reaches its designated discharge point, the tray drops open like the bomb-bay doors of a bomber plane and drops the product into a tote or chute. Like the pusher tray, this is used for soft, small, or irregularly shaped products (Figure 6.30).

6.3.9.5 Chutes

Many of these sorters divert the product into chutes. The chutes can also aid in the sorting process. In Figure 6.31, we have a chute that has a secondary divert built into it. Notice that the first two chutes seem to drop down underneath the chute surface. The third chute is straight. This is because these chutes have two destinations, one above the other.

These chutes can be designed to sort up and down or left and right. There typically are only two destinations per chute.

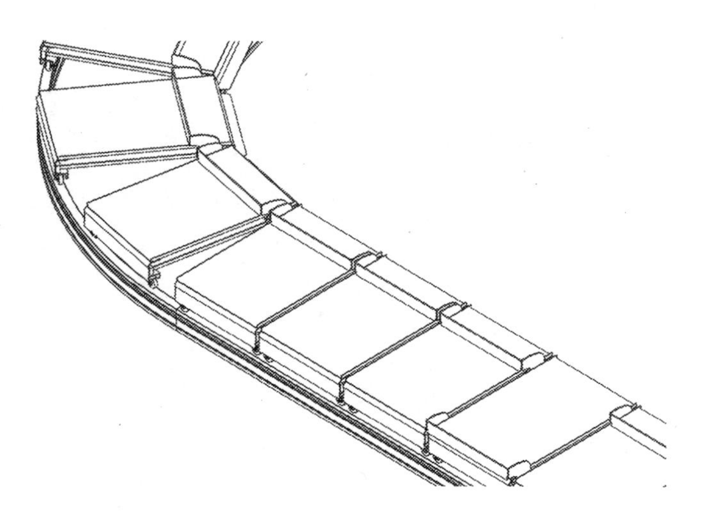

FIGURE 6.29 Pusher Tray sorter.

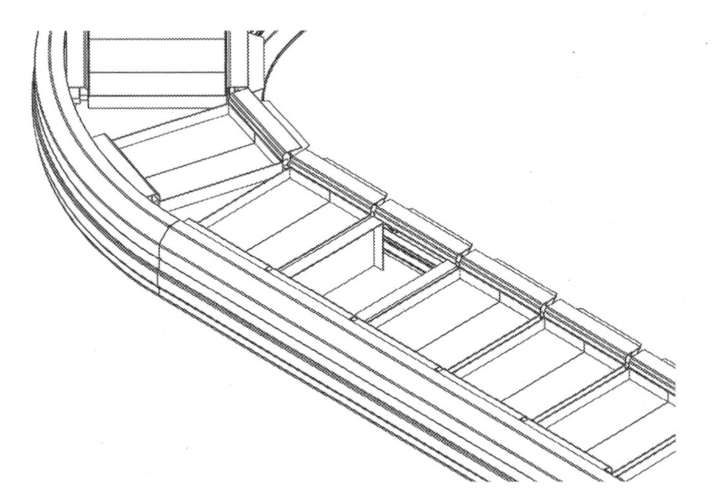

FIGURE 6.30 Bomb-bay sorter.

FIGURE 6.31 Dual-level chute.

6.4 QUESTIONS

1. What are the advantages of a small end pulley on a belt-driven live roller conveyor?
2. What are the disadvantages of a small end pulley on a belt-driven live roller conveyor?
3. With a pusher or transfer sorters what is one important issue specific to them?
   a. Spacing between the products.
   b. After the divert, the product is traveling in a different orientation.
   c. The take-away conveyor has room for more products.
   d. The take-away conveyor is running fast enough.
4. What is the longest 24″ wide box that will properly convey through a curve that has a 36 inside radius and 28 between frames?
5. What product characteristics do you need to consider when selecting a bomb-bay sorter?
   a. Is the product fragile?
   b. Will the product leak depending on orientation?
   c. Will dropping other products damage the packaging?
   d. All of the above.
6. Which of the following is best suited for sorting folded apparel in bags?
   a. Pop-up transfer sorter
   b. Tilt tray sorter
   c. Sliding shoe sorter
   d. Bomb-bay sorter
   e. b and d
7. Given that a sliding shoe sorter is best for handling bags of dog food, what types of conveyors should be used for the balance of the system?
8. Given boxes that range from a minimum size of 12″ W x 18″ L up to a maximum size of 24″ W x 30″ L, how long of a skewed roller area do you need to get all boxes to one side of a 30″ BF conveyor?
9. Given the average person can induct 30 products per minute on a push tray sorter and the sorter can run 18,000 trays per hour, how many people are required to induct product to achieve 150 sorts per minute?
10. Beyond functionality, when adding to an existing conveyor system, which of the following plays into the decision of what conveyor to use?
    a. Use the same brand and type conveyor as exists to minimize new spare parts.
    b. Use similar type of conveyor as exists because it's already proven.
    c. Use similar type of conveyor as exists because the maintenance people are familiar with it.
    d. Use something different because the customer has had a bad experience with the existing conveyors.
    e. All of the above.

7 Heavy Unit Load Handling Conveyors

This chapter covers a wide range of conveyors used for moving heavy unit loads, defined as any single item being handled weighing more than 200 lb. This can include drums, crates, slip sheets, and pallets carrying various products, be it cases of wine, bags of dog food, or partially assembled diesel engines. Among heavy unit loads, pallets make up the vast majority of applications, and the majority of pallets are made of wood. For that reason, the words *pallet* and *product* are used interchangeably in this chapter.

We first review the technical aspects of the various heavy unit load handling (HULH) conveyors and then discuss how best to apply the conveyors.

7.1 GRAVITY ROLLER CONVEYORS

HULH gravity roller conveyors are, for all intents and purposes, the older sibling to the smaller gravity roller for packages. The side frames are typically either made from formed $^{3}/_{16}$-in. or ¼-in. steel or made from structural steel channel. The rollers are around 2½ in. or larger in diameter, with wall thicknesses of 11 gauge or greater.

The key in applying gravity rollers for pallets is to set them at a pitch that keeps the pallets moving at a set speed. Keep in mind that the pitch to keep a poor-quality pallet moving might be so steep as to allow a good-quality pallet to move too quickly. A heavy load moving on gravity rollers and then stopping abruptly can be dangerous, or even deadly if the load topples.

7.2 ROLLER TRANSPORTATION CONVEYORS

Just like the lighter package conveyors, there are a variety of roller transportation conveyors. Some of them are beefed-up versions of their light-duty versions. The different types of roller drives have a direct impact on the weight that the conveyor can move. CDLRs have the greatest load capacity at more than 4,500 kg (10,000 lb.). Belt-driven live rollers top out at 4,000 lb., the lineshaft-driven live roller capacity is 225 kg (500 lb.), and padded chain-driven live rollers have a maximum capacity of 900 kg (2,000 lb.).

The primary transportation tool in HULH is CDLRs. There are two types of CDLRs (see Figure 7.1). The first uses continuous chain to drive multiple rollers. This is considered a lighter duty type of CDLR because there are fewer sprocket teeth engaged on each roller. The second and by far the most popular is the roller-to-roller type in which chains connect each roller to the adjacent rollers. This is what most people immediately think of when you mention CDLRs. In both cases, the rollers are driven by the sprockets that are welded to one or both ends of the roller. Just like with package conveyors, there are several other types of HULH live roller conveyors. Belt-driven live roller conveyors, lineshaft conveyors, and padded chain-driven live roller conveyors are discussed.

DOI: 10.1201/9781003376613-7

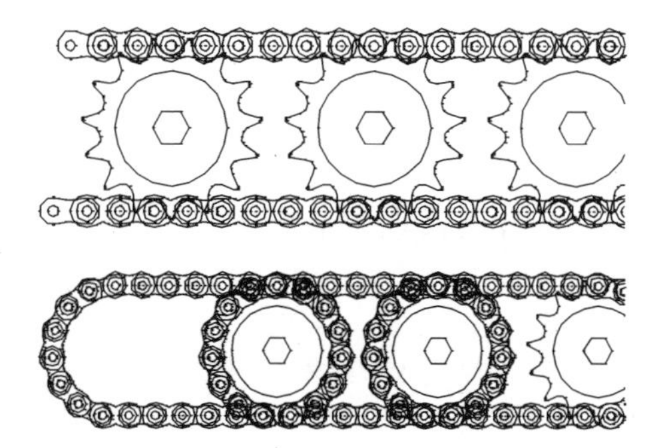

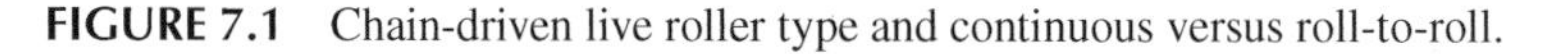

FIGURE 7.1 Chain-driven live roller type and continuous versus roll-to-roll.

7.2.1 Continuous Chain CDLR

The continuous chain type of CDLR is typically used for what would be considered medium unit loads, between 100 and 500 lb. As mentioned earlier, the rollers are driven by one or two chains that engage the rollers from the top and bottom. This drive capacity is limited to the strength of the chain.

Most frequently, #50 roller chain is used to drive the rollers. The chain can be a single strand, two strands, or a single strand of double chain. The rollers are most frequently a heavy-walled 1.9-in. diameter.

Typically, the chain is held against the sprockets with the use of chain guide strips. These guides are made from either cold-rolled steel or ultra-high-molecular-weight polyethylene (UHMW).

The drive for continuous chain CDLR is typically mounted under the conveyor. Figure 7.2 shows a center drive. Note that the drive has the capability to take up any slack in the drive chain. This is important when using continuous chain CDLR because over time the drive chain will stretch.

The products that can be handled on continuous chain CDLRs, like all roller conveyors, are limited by the roller spacing. The minimum spacing is determined by the outside diameter of the sprockets that are welded to the rollers.

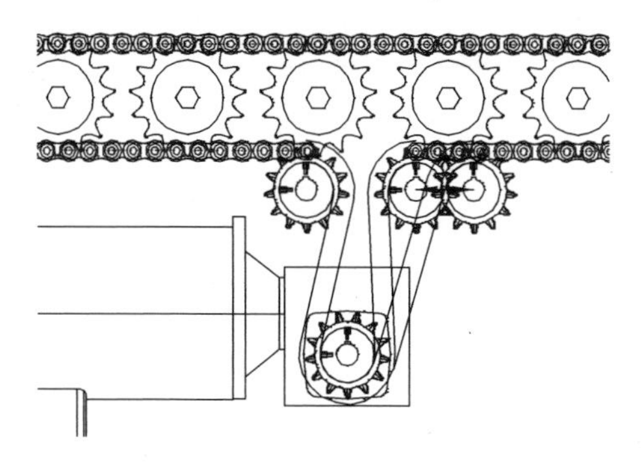

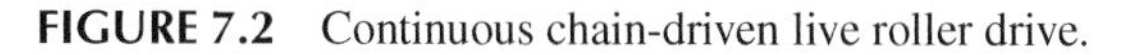

FIGURE 7.2 Continuous chain-driven live roller drive.

7.2.2 Roll-to-Roll CDLR

As mentioned earlier in this chapter, roll-to-roll CDLR is the most popular and is used to carry much heavier loads. This type of CDLR handles truly heavy unit loads of anything more than 200 lb.

Like all roller conveyors, the products that can be handled on roll-to-roll CDLR are limited by the roller spacing. The minimum spacing is determined by the outside diameter of the sprockets that are welded to the 2½-in. or $2^9/_{16}$-in. diameter rollers. Table 7.1 is a chart of roller centers versus sprocket size. Every manufacturer has its own standards, but these are fairly common.

Table 7.1 shows only #40 through #60 roller chain, but for heavier loads #80 or larger chains can be used. Typically, when larger chains are used the rollers grow to 3½ in. in diameter or larger.

There are ways to compress the roller further. Figure 7.3 shows some variations on the common theme. Figure 7.3a shows an offset drive concept. This has two complete sets of roll-to-roll chains; each drives every other roller. Figure 7.3b shows the rollers being driven by sprockets on the roller shaft. Here, rather than fixed shafts and rollers having the typical bearings in the ends of the roller, the bearings are on the frame and the roller and shaft both rotate.

To minimize roller centers, some manufacturers offer CDLR with every other roller driven, as shown in Figure 7.3c. The nondriven rollers are static gravity rollers. I strongly discourage the use of this design because the driven roller next to a free-spinning static roller creates a pinch point like a wringer washing machine scenario, which is very dangerous for nearby personnel.

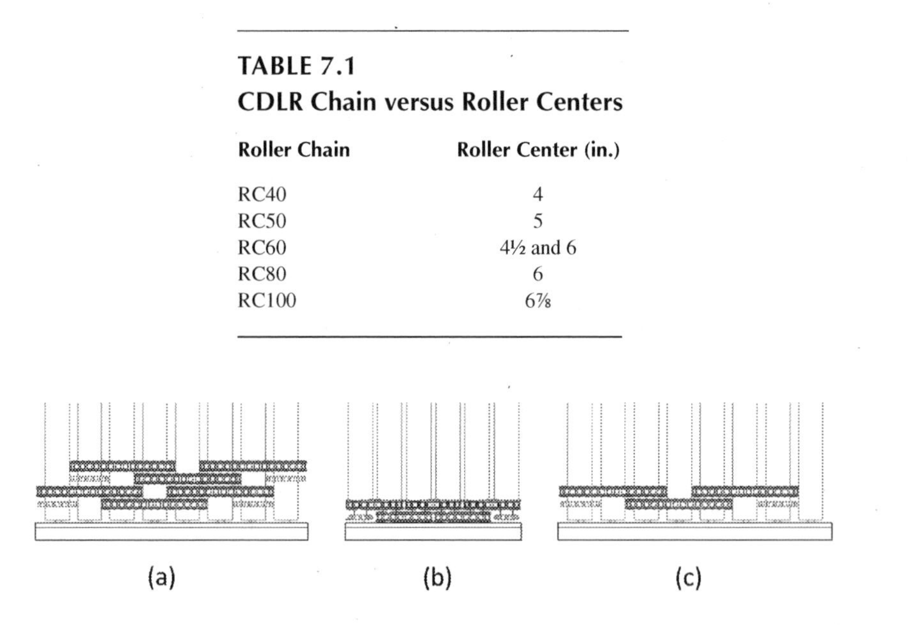

TABLE 7.1
CDLR Chain versus Roller Centers

Roller Chain	Roller Center (in.)
RC40	4
RC50	5
RC60	4½ and 6
RC80	6
RC100	6⅞

FIGURE 7.3 Alternative drive chain arrangements: (a) offset drive, (b) shaft-mounted sprockets, and (c) every-other-roller driven.

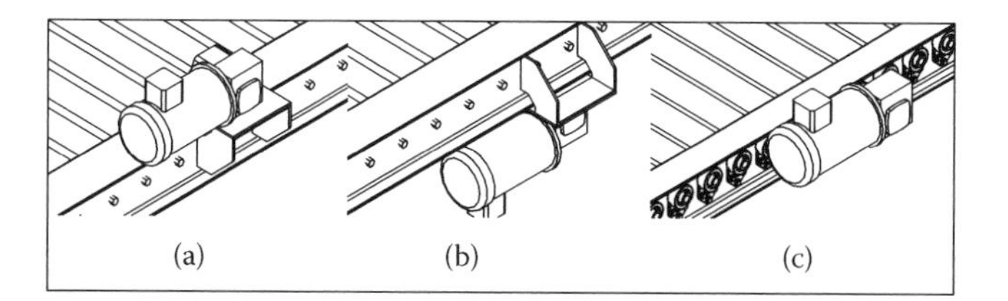

FIGURE 7.4 Drive mounting options: (a) top-mounted drive, (b) underhung drive, and (c) shaft-mounted drive.

Roll-to-roll CDLR can be driven in any of three ways. The most common is the top-mounted drive, as shown in Figure 7.4a. Figure 7.4b shows a bottom-mounted or underhung drive, which requires that the conveyor sit higher off of the floor. It is important to keep in mind that there must be sufficient room to remove the chain guard. This typically requires an additional 12 in. or more below the guard. Both the top-mounted and the underhung drive mounts must have a method of adjustment to properly tension the drive chains. The third type of drive is the shaft-mounted drive. Figure 7.4c shows that the roller on which the drive is mounted is on bearings that are mounted on the side frames. Although the shaft-mounted drive costs more initially, it is the easiest to maintain.

Just like the lighter weight roller conveyor, there is motorized roller CDLR. The motors can be 24V or 48V DC powered or 120V single phase or 230V 3 phase powered. These come with the sprockets welded on from the manufacturer. Due to the size of the motor, they are limited in how heavy a load they can reliably convey.

The horsepower to drive a CDLR is dependent on both the unit load to be moved and the length of the conveyor. As with any chain drive, there is a loss of drive capacity due to the chain. Therefore, each chain loop will decrease the power transmitted to the next roller as they get further from the drive. This is why the drive is typically located toward the center of the conveyor.

7.2.3 BDLR

The heavy unit load BDLR is very similar to the lighter duty version used for packages. The difference is that the rollers and belt are both larger. The belt is a minimum of 12-in. wide and is typically no wider than 18 in. You will not find any v-belt or round-belt-driven HULH conveyors. Due to the inherent flex of the carry rollers as a heavy load goes over, the pressure rollers are typically equipped with a spring mechanism to keep the rollers pressed against the bottom of the carry rollers while offering forgiveness when the carry roller is pushed down.

With BDLR, because there is nothing added to the outside of the rollers, they can be placed close together. Roller centers for BDLRs can be 3, 4, 6, or even 9 in. Obviously, the closer the rollers are, the more weight they can handle but also the more expensive is the conveyor. Because there is an increased likelihood of a person's hand, feet, or clothing slipping between the rollers, many manufacturers offer a pop-out roller feature. The hex shaft or the rollers are in slots that extend up through the top of the frame; this allows the roller to lift up if something gets underneath.

If the frame does not have this pop-out feature, for safety reasons, we strongly recommend against using roller centers more than 3 in.

7.2.4 Padded Chain-Driven Live Roller

Just as with the other conveyors we have discussed, heavy unit load padded chain-driven live rollers are very similar to the lighter duty version. The rollers are a larger 2½ in. diameter, but the same type of roller chain with urethane pads snapped into it is used to drive the rollers. The chain rides on a UHMW track that is supported on springs so that the chain is always in contact with the bottom of the carry rollers.

To increase drive capacity, this conveyor can be manufactured with two separate drive chains driven by the same motor. This type of conveyor can carry loads of up to 1,200 lb. with a single chain and 2,000 lb. with dual chains.

7.2.5 Lineshaft

Heavy unit load lineshaft conveyors use the same theory as the lighter duty version. The rollers, shaft, spools, and urethane bands are all larger, but the mechanics of the conveyor are identical. Typically, instead of a 1-in. shaft and $^{3}/_{16}$-in. urethane bands, they are beefed up to $1^{7}/_{16}$-in. shafts and ¼-in. urethane bands. This greatly increases the driving capacity of the conveyor. HULH lineshaft conveyors have adequate drive for loads up to 500 lb.

7.3 ROLLER ACCUMULATION CONVEYORS

HULH accumulation conveyors are typically a variation of CDLRs, but some manufacturers offer a BDLR accumulation.

7.3.1 CDLR

CDLR accumulation is basically a series of CDLR sections that are driven by pneumatic clutch brakes.

The accumulation is typically controlled by photo-eyes or a combination photo-eye and solenoid valve. A solenoid valve is typically used to arm the first zone in a conveyor. The next product triggers the sensor (photo-eye or photo-valve), the clutch is disengaged, and the brake is engaged. This stops the rollers, and the preceding zone is armed. The next product triggers the sensor, the clutch is disengaged, and the brake is engaged to stop the rollers. This scenario is repeated again and again until all of the zones are full.

The horsepower to drive a CDLR accumulation is dependent not simply on how long the conveyor is but also on the maximum number of unit loads that will be in motion at any given time. It is not necessary to assume all loads will be in motion at once, unless that is how the conveyor is to be operated (Figure 7.5).

All of the rollers had to be replaced with new rollers that were 1/8″ shorter.

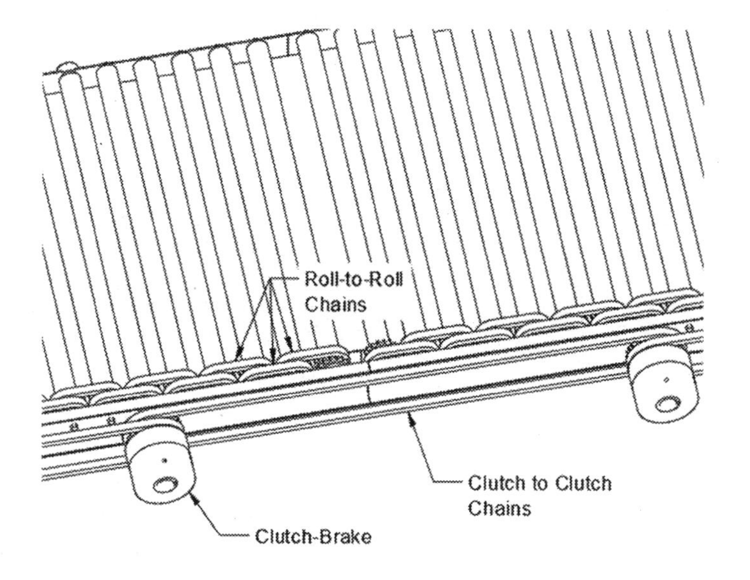

FIGURE 7.5 CDLR accumulation conveyor.

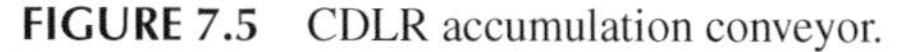

CASE STUDY – ROLLER "GROWTHS"

I was called in to solve a problem with a series of CDLR accumulation conveyors. The system was 8 or 10 conveyors set tightly side by side. They were loaded and unloaded by AGVs (automatic guided vehicles). The product was a heavy-duty wire cage carrying crankshafts that have just come from a furnace. The bottom of the wire cage was cast Ni-hard steel. The problem was it appeared as though steel residue was building up on the surface of many of the rollers. The customer's theory was that the hardened steel bottom of the cages was wearing steel off some rollers and depositing it on other rollers causing growths.

I watched the system run for several hours and saw nothing to support the customer's theory. We took a new spare roller and replaced one of the worst rollers so that we could examine it. Figure 7.6 shows what we found when we cut it open. The roller had actually wrinkled either side of center. I gathered all of the best, most experienced engineers and product managers that Rapistan had at their main facility. Together, we boasted over 200 years of collective experience. No one had ever seen anything like this. After reviewing the application thoroughly including pictures and video we were no further ahead. We tried one of our automotive customer's problem-solving process, which was a process of eliminating everything that was NOT the cause. This provided for some levity when it was pointed out the sun shining was NOT a cause. After several hours of eliminating more relevant conditions, it was finally determined that the rollers were too long for the conveyor frame.

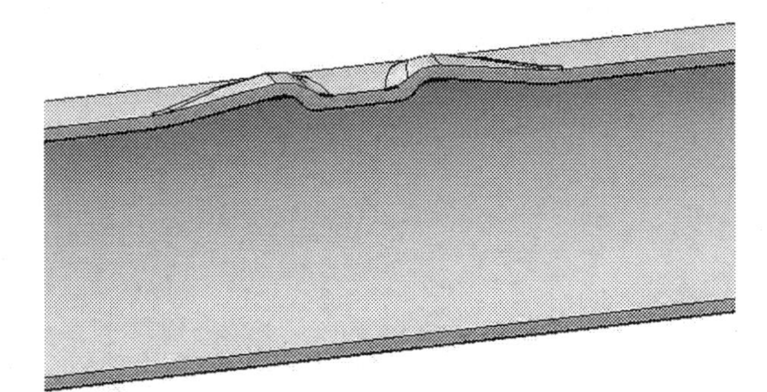

FIGURE 7.6 Cross-section of damaged roller.

The overall dimension from bearing to bearing of a roller is normally 1/8″ less than the between-frame dimension. As we have mentioned earlier, all rollers are slightly banana shaped. No roller is perfectly straight. In a normal conveyor if a heavy load is on a roller when the roller is arched up, the roller needs to flex outward. Typically, this is not an issue because there is 1/8″ of space for it to do so. In this case, the rollers were not 1/8″ less, but the same dimension. This did not allow the roller any room to move or expand. All the conveyors were installed side by side, right against each other so even the side frame could not deform to allow room.

In this case, there was no room and so when the center rib pushed down on the high side of the roller, the only thing that could happen was the material buckled or wrinkled up either side of the center rib of the cage's Ni-hard bottom frame.

7.3.2 BDLR

The BDLR, like the lighter duty version, uses pneumatic actuators to keep the drive belt pressed against the bottom of the conveyor rollers. Typically, the zones are 3- to 5-ft. long, depending on the products being conveyed. Unlike CDLR accumulation, BDLR can also utilize sensor rollers rather than just photo-eyes or photo-valves to control the accumulation logic.

The accumulation uses the same type of logic as the light-duty conveyor. A solenoid valve is typically used to control the first zone in a conveyor. When the valve is triggered, the drive belt drops away, removing the drive from the rollers. The next product then coasts into the zone and triggers the sensor (roller or photo-eye). This in turn removes the drive from the preceding zone. The next product coasts into that zone and triggers the sensor, and the scenario is repeated again and again until all of the zones are full.

7.3.3 Padded Chain-Driven Live Roller

The accumulating version of the padded chain-driven live roller conveyor uses heavy-duty sensor rollers to control the pneumatic logic of the accumulation action. Like BDLR, the accumulation uses the same type of logic as the light-duty conveyor. A solenoid valve is typically used to control the first zone in a conveyor. When the valve is triggered, the drive belt drops away, removing drive from the rollers. The next product then coasts into the zone and triggers the sensor (roller or photo-eye). This in turn removes the drive from the preceding zone. The next product coasts into that zone and triggers the sensor, and the scenario is repeated again and again until all of the zones are full.

7.3.4 Lineshaft

Lineshaft accumulation uses heavy-duty brakes to stop the rollers, which are actuated using either photo-eyes or photo-valves. The way lineshaft accumulation conveyors operate is very similar to CDLR accumulation. A solenoid valve is typically used to arm the first zone in a conveyor. The next product triggers the sensor (photo-eye or photo-valve), the brake is engaged, the rollers are stopped, and the preceding zone is armed. The next product triggers the sensor, the brake is engaged to stop the rollers, and the scenario is repeated again and again until all of the zones are full.

7.3.5 Accessories

The following common accessories are used in conjunction with HULH roller conveyors:

- Flanged rollers: Flanges are added to the rollers of CDLR as a method to guide the product. Opposing flanges can be added to every other roller to guide the product without unnecessarily increasing the roller centers. This guiding method is very popular in Europe where pallet sizes and construction are more consistent.
- Product guides: To guide the product on roller conveyors, the preferred method is to use wheels rather than a guide rail. The drag created by a guide rail is often more than the drive of the conveyor can overcome.
- Product stops: These are used to accurately stop a product.
- Squaring devices: These can be a fixed heavy-duty end stop that the product is driven into to force it square with the conveyor, or it can be a squeeze-type device that pushes the product sideways on the conveyor to square it with the conveyor.
- Roller brakes: Brakes are used not only on lineshaft conveyors but also on BDLRs and padded chain-driven live rollers to accurately stop a product. This is used when simply removing the drive from below the rollers is not sufficient to stop the product.

7.4 MULTISTRAND CHAIN CONVEYOR

Multistrand chain conveyor is sometimes also referred to as drag chain conveyor. The principle is that a series of roller chains act as the conveying surface. The chain is typically RC60, RC80, or even larger for pallet conveyors and RC50 and smaller for lighter duty applications. There are specialty chains for this use as well for certain applications, such as handling lumber.

One of the keys to chain conveyors is that the chain must be straight. That sounds obvious, but most people are unaware that standard, off-the-shelf roller chain has some inherent twist in it. This twist will make the chain fall off of the chain guides and get bound at the sprockets. We recommend using straightened chain. Most chain manufacturers offer straightened chain as an option.

Multistrand chain conveyors have multiple strands of roller chain (see Figure 7.7). There can be from two to as many strands as necessary to support and pull the load. Of course, the more strands that are used, the more expensive the conveyor becomes. The size and number of chains are determined by the loads to be conveyed.

The simplest drive for a chain conveyor is an end drive; however, the sprocket might be too large for the product to transition to the next conveyor smoothly. End drives are not reversible either. These are the two reasons many chain conveyors have a center drive. This allows small idler sprockets on the ends and bidirectional operation.

Chain conveyors are used primarily when the bottom boards of a pallet are perpendicular to the direction of the conveyor. This is an ideal application for chain conveyors, but the ends of the conveyors should keep the pallet supported as it transitions from one unit to the next so that the bottom boards do not drop between the conveyors. This can be done by overlapping the end of the conveyors or by adding some static rollers on either side of the chains to keep the pallet supported.

A special type of multistrand chain conveyor is frequently referred to as a strapper conveyor. This conveyor is typically designed with five strands of chain and an integral turntable (see Figure 7.8). The idea of the conveyor is that as a pallet load of product exits a bulk palletizer, discussed later, the conveyor stops so that a strapper

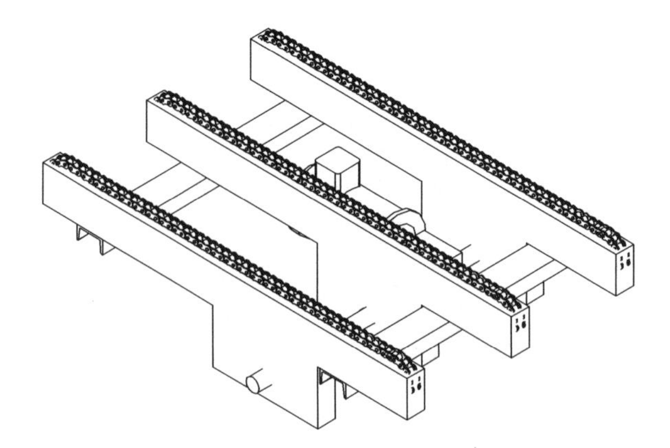

FIGURE 7.7 Three-strand chain conveyor.

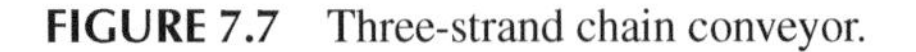

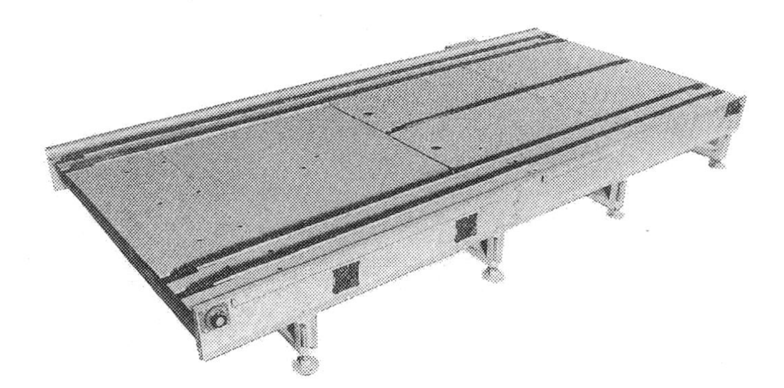

FIGURE 7.8 Specialized strapper conveyor.

can put a strap around the pallet and the load that is on it. The load is moved forward so that multiple straps can be applied. When all of the straps have been applied, the pallet moves forward to the turntable, is lifted and rotated 90 degrees, and then set back down on the conveyor chains. The pallet then is conveyed back through the strapper for additional straps. When the process is complete, the pallet can exit the conveyor or go back to the turntable and be reoriented again and then exit the conveyor.

7.4.1 Accessories

The common accessories that are used in conjunction with HULH multistrand chain conveyors are the following:

- Product guides: Fixed angle guide rails are frequently used on chain conveyors. The conveyor has ample drive to overcome any drag caused by the product coming in contact with the rail. Fixed guides are less expensive than guide wheels that are used on roller conveyors.
- Product stops: These are used to accurately stop a product.
- Squaring devices: These can be a fixed heavy-duty end stop that the product is driven into to force it square with the conveyor or it can be a squeeze-type device that pushes the product sideways on the conveyor to square it with the conveyor. These are not recommended if the product is more than 5 degrees out of square. That is because as the product twists back in line with the device, it can push the chains off the guides.

7.5 RIGHT-ANGLE TRANSFERS

There are three ways to change direction with HULH conveyors. The first is a CDLR curve, and the second is a right-angle transfer. Whenever applying a transfer, keep in mind that the orientation of the product as it relates to its direction of travel changes. There are two primary types of right-angle transfers. For CDLR there are pop-up chain transfers and for multistrand chain conveyors there are pop-up roller transfers

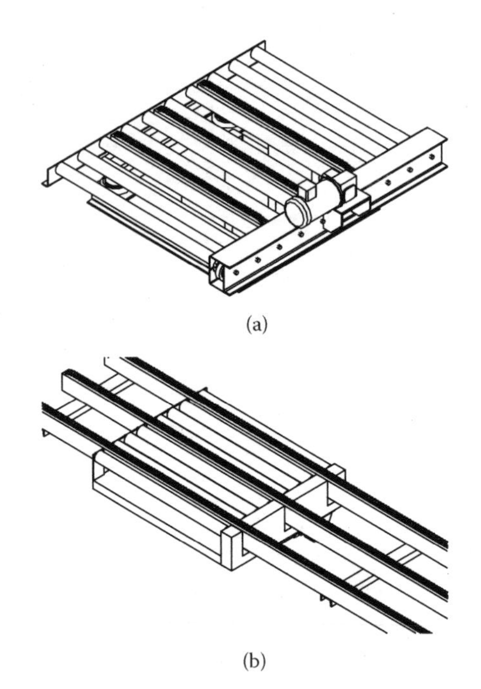

(a)

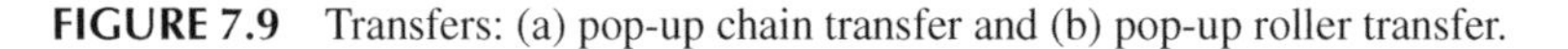

(b)

FIGURE 7.9 Transfers: (a) pop-up chain transfer and (b) pop-up roller transfer.

(see Figure 7.9). The concept is that the transfers sit below the normal conveying surface. When a product is in the correct place, the transfer is raised, and the product is lifted off the first conveyor and carried off at a 90-degree angle. The conveyor that the product is transferred onto must be designed to handle the product in its new orientation.

The vast majority of transfers have their own drive to power the chains or rollers. The only exception we have seen is on heavy-duty lineshaft conveyors. One manufacturer offers a slave-driven transfer, but this is the exception and is limited in its capacity.

When applying a transfer, it is important to stop the product *before* raising the transfer. This is generally considered good practice, but it is vitally important with pop-up chain transfers because the moving product can push the chains off the guides and cause serious damage. For pop-up roller transfers, it is not as important; the primary concern is premature wear of the lift mechanism and guides.

With pop-up chain transfers, it is also important to ensure that the product stops over the chains of the transfer. Often additional chains are used in a transfer, which allows multiple product sizes to be lifted and conveyed reliably by a transfer.

There are several different types of lifting mechanisms. Table 7.2 presents a list of the most common and Figure 7.10 shows them; each has advantages and disadvantages.

Lifting capacity for each design is not listed because each manufacturer has its own design, and the geometry of the mechanism can have a dramatic effect on capacity.

TABLE 7.2
Transfer Lift Types

Lift Type	Advantages	Disadvantages
Air bags	• Less expensive • Simple to apply	• Lifting can be uneven • Requires vertical lift guides • Minimum four items to maintain, replace, or both
Fire hose	• Least expensive • Simple to apply	• Lifting can be uneven
Air or hydraulic cylinders powering a crank lever	• Smooth parallel lift • One primary maintenance item	• More expensive than air bags or fire hose • Bearings or bushings that need lubrication and maintenance
Air or hydraulic cylinders pushing a lifting wedge	• Smooth parallel lift • One primary maintenance item	• More expensive than air bags or fire hose • Bearings or bushings that need lubrication and maintenance
Electric linear actuator	• Smooth parallel lift • One primary maintenance item • Does not require compressed air or hydraulic pump	• More expensive than air bags or fire hose

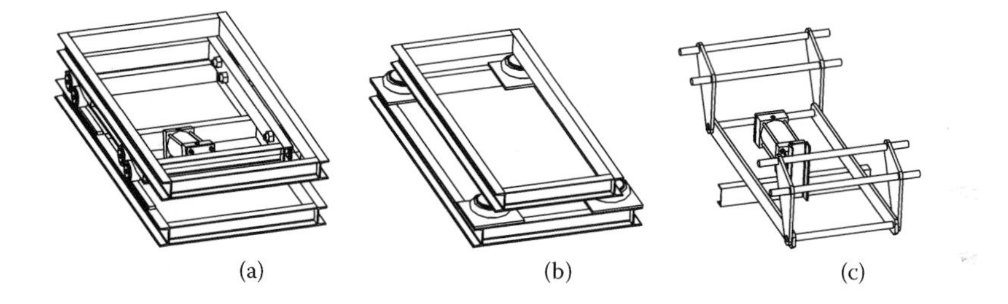

FIGURE 7.10 Lift mechanisms: (a) cylinder powered wedge lift, (b) air bag lift, (c) cylinder powered crank lift.

Any time air is used for a lifting device, it is important to remember that air is compressible. That means when more weight is added, the transfer might not lift as fast at the same air pressure. In the case of air bags, due to having multiple lifting points, an uneven load can cause the lift to come up skewed and possibly jam if the transfer is not properly guided.

Notice that what would seem an obvious choice, four air cylinders, is not listed. The reason for that is multiple air cylinders never actuate exactly together. Each cylinder has a different breakaway pressure, the pressure at which a cylinder starts to move. This causes the cylinders to actuate independently, and when one is lifting against the others, it can get jammed or twist the cylinder rod. Experience teaches that no matter how hard you try, you will never get four cylinders to actuate exactly together.

One entry that catches a lot of people by surprise is the fire hose. This is applied by using a section of fire hose clamped shut at both ends under each strand of chain in a chain transfer. The fire hose is inflated and just the chains and product lift. The drive mechanism for the chains does not have to move. This is frequently used in some of the lighter heavy unit loads in the 90–450 kg (200–1,000 lb.) load range.

Compressed air is typically readily available in most industrial facilities, so air-operated lifting seems the most logical choice. Hydraulics, although more expensive, offer the most consistent actuation. The cost of the cylinder is negligible. Air cylinders are less expensive, but due to the higher operating pressure, a smaller hydraulic cylinder can be used. The added cost comes from the need for a hydraulic pump, oil reservoir, and high-pressure lines.

Some applications cannot tolerate even the remote possibility of an oil leak, such as in a food plant, so air is the only choice.

Transfers, although offering system flexibility, can have an adverse effect on rate because the product rate through a transfer area is lower than the normal straight-line rate. The rate calculation is discussed in Chapter 10, "Rate Calculations." The calculation determines product spacing that would allow the system to operate without stopping the pallets except to position them on the transfer. However, it is strongly recommended that the system be designed so that there is an opportunity to stop oncoming product prior to the transfer. This ensures that there is no chance of a product running into the raised transfer.

7.6 TURNTABLES

The third way to change directions on HULH conveyor systems is through the use of a turntable. Like curves, turntables keep the product in the same orientation as it relates to the direction of travel.

A turntable is a device that has a CDLR or multistrand chain conveyor mounted on top of it (see Figure 7.11). Turntables typically have two drives: one to operate the conveyor on top and a second to turn the turntable.

The conveyor on a turntable has to adhere to the same design criteria applied to the rest of the conveyors. The turntable itself, as mentioned, is usually motor driven, although some manufacturers offer air or hydraulic-actuated turntables as well as manual push turntables for lighter loads. Air is not recommended except for the very

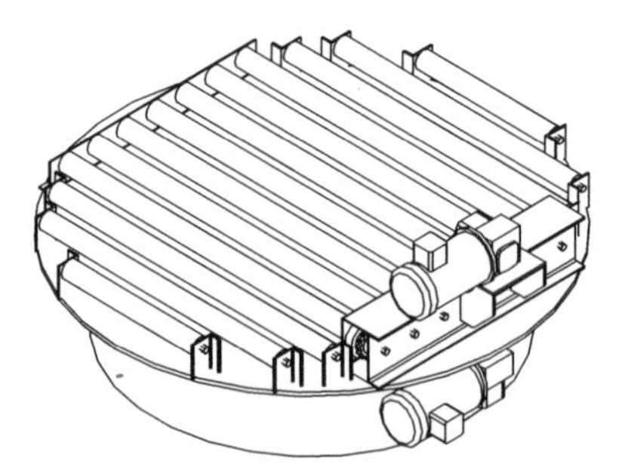

FIGURE 7.11 CDLR on a turntable.

lightest loads; the reason is that air is compressible and the turntable will not turn evenly and smoothly but will jerk and bounce as it turns.

Turntables are capable of handling loads up to 445 kg (10,000 lb.), depending on the turning mechanism and the support system. Underneath every turntable is some sort of support structure. The most common structure uses four heavy-duty casters to support the turntable. Some manufacturers offer a ring of wheels and still others use very large radial bearings to support the turntable.

Turntables are a great alternative to CDLR curves when space is limited. Turntables can also handle long, narrow product that would otherwise require a very large CDLR curve. They are also the only choice when using multistrand chain conveyor that allows you to keep the product in the same orientation as it relates to the direction of travel.

Because the ends of the chain-driven live roller (CDLR) conveyor on the turntable are tapered, there is clearance for turning. Something is needed to support the edges of the product as it enters and exits the turntable. Another important aspect of the design is how many of the rollers are driven on the ends. This becomes important when smaller product is being handled. Between the conveyor feeding it and the turntable itself, there must be enough powered rollers under the product at all times to keep it moving reliably.

Most powered turntables can handle up to three products per minute. That is how long it takes for a product to convey on, turn, convey off, and then for the empty turntable to turn back into position for the next product.

Turntables can have an adverse effect on rate as compared to the flow through a curve, as discussed in Chapter 10. The calculation determines product spacing that would allow the system to operate without stopping the pallets except to position them on the turntable. However, the system should be designed so that there is the opportunity to stop oncoming product prior to the turntable. This ensures that there is no chance of a product running into a turntable that is not in position.

7.7 ANCILLARY EQUIPMENT

We very briefly touch on some of the ancillary equipment that is typically seen when dealing with HULH equipment. The information is intentionally brief to give an overview of the equipment and its purpose.

7.7.1 Lifts

When handling heavy unit loads, there is no good way to change elevations using an incline conveyor, unless the elevation change is slight. The maximum recommended incline is 5 degrees. This can require a great deal of distance to climb even 1 ft. The preferred method to change elevation is to use a vertical lift. There are several different types of lifts, and each has its strengths and weaknesses, as listed in Table 7.3.

Vertical reciprocating conveyors (VRC) come in a variety of designs, including straddle, cantilevered, and four-post designs. Each offers features that can be useful depending on the application. Each manufacturer of these lifts typically offers several different configurations to take advantage of those features.

TABLE 7.3
HULH Lift Types

Lift Type	Strength	Weakness
Scissor lift	Maximum load: 20,000 lb.	Limited vertical travel distance
Vertical reciprocating lift, air operated	Maximum load: 1,000 lb.	Rate is limited because the platform must return between pallets Required high volume of plant air
Vertical reciprocating lift, hydraulic operated	Maximum load: 4,000 lb.	Rate is limited because the platform must return between pallets
Vertical reciprocating lift, motor driven	Maximum load: 30,000 lb. Height change: 100 ft.	Rate is limited because the platform must return between pallets
Continuous vertical lift	Maximum load: 5,000 lb. Higher, continuous throughput	Lower weight capacity

Note that VRCs are never referred to as elevators because this opens a litany of regulatory concerns. VRCs are not intended for a person to ride with the product unless it is specific designed as such by the manufacturer.

7.7.2 Palletizers

Palletizers are a class of machine that is used to stack or palletize product. Product such as boxes or bottles are stacked on a pallet as a way of transporting large quantities as a single load. There are three primary types of palletizers: bulk, carton, and bag.

Bulk palletizers are used to palletize individual bottles or cans. The bottles or cans are stacked in layers with a thin piece of cardboard, slip sheet, or tier sheet between the layers. The stacked load can be up to 8-ft. tall, the typical maximum height that can fit into a tractor trailer. The load is topped off by a "picture fame" or "top frame." A top frame is either a rectangular frame made of wood with another tier sheet attached to it or, in more recent years, a blow-molded plastic frame. This provides structure and stability to the palletized load. Depending on the product being palletized, loads can reach weights of up to 5,000 lb. as is the case with 2-oz. glass baby food jars. Upon exiting the palletizer, bulk loads are always strapped to the pallet by a strapping machine that typically uses plastic strapping material. The strapping process compresses the load and adds to its structural integrity.

Carton palletizers stack multiple layers of boxes or cartons onto a pallet. Each layer has a predetermined pattern, and the pattern is typically alternated between layers so that the boxes are interlocked to provide stability (see Figure 7.19). The infeed of the palletizer has a device that will turn certain boxes, so that as they come into the machine, the pattern is built. Carton palletizers typically do not use a tier sheet between layers.

Bag palletizers work almost identically to carton palletizers with alternating patterns to each layer. The primary difference is that most bag palletizers are preceded by a bag flattener. This makes each bag a consistent thickness so that the load is built evenly.

With the significant cost reduction of industrial robots in recent years, there are many robotic palletizing operations. These are more flexible than traditional palletizers and can be reconfigured. A single robot can be set up to palletize products from different conveyor lines at the same time. The robot can handle individual boxes, a row of boxes, or an entire layer of boxes at a time. It's all dependent on rates and weights.

7.7.3 Pallet Handling Accessories

Most palletizers are equipped with a pallet dispenser, a machine that takes a stack of pallets and dispenses them onto the conveyor one at a time. As with any equipment, there are several different types. Some use a chain conveyor underneath the stack with large dogs mounted to the chains that strip the bottom pallet out from under the stack. This type is the least expensive but is the most abusive to the pallets. Others have a variety of mechanisms to lift the stack off of the bottom pallet, allowing it to be conveyed out without the resistance of being dragged out from under the stack. Most pallet dispensers are designed for a specific size of pallet. Some are built to be adjustable, but all are designed to handle only one pallet size at a time.

Pallet accumulators are very similar to pallet dispensers. They are used to recreate the stacks that the pallet dispenser took down.

Pallet dispensers can be loaded either by a forklift or by an infeed conveyor that can have multiple stacks of pallets waiting to be dispensed. Pallet accumulators can be unloaded using the same concepts; either a forklift or a conveyor carries away the completed stack.

Pallet inverters are designed for transferring various palletized products from one pallet to another without manually restacking the load. Each load inverter is built with a side plate for supporting the load during rotation. The load is clamped between two platens, rotated 180 degrees, and then the lower platen is released so the pallet can be removed from the load. The new pallet is replaced by another, and the procedure is reversed. This can also be used to transfer loads to slip sheets or to replace damaged product on the bottom of a palletized load without restacking the entire load.

The transfer car is a fixed-path conveyor, just like everything else we have discussed in this book. Automatic guided vehicles (AGVs) do not follow a fixed path but follow programmable paths. Transfer cars have either one or two CDLRs or multi-strand chain conveyors mounted on them. The cart follows tracks that are mounted on the floor. This allows the area where a fixed conveyor would otherwise be to be open to transport and pedestrian traffic. Transfer cars typically have to be equipped with many of the same safety features as an AGV to prevent collisions with objects in the car's path. There are three primary ways to get power and control signals to the transfer car: cable chain, retractable cable reel, and a festoon system. Each and every transfer car is custom designed for the specific application, including the track type, the drive, and the control system.

AGVs and AMRs (autonomous mobile robots) do not follow a track. Instead, they have a variety of guidance options. Older technology used a wire in the floor that the AGVs followed. They then progressed to reflector strategically located along their path. With the advent of Lidar (light detection and ranging), AGVs and AMRs can follow any path that they are programmed with. The Lidar allows them to map their stationary surroundings as well as scan for obstacles such as pedestrians.

7.8 APPLICATION DETAILS

As part of discussing the application of HULH conveyors, it is important to first discuss the product to be handled. As mentioned at the beginning of this chapter, pallets make up the vast majority of heavy unit loads and that the majority of pallets are wooden.

When analyzing HULH applications, it is important to divide the overall handling requirements into operational areas such as receiving, staging, warehouse storage, sorting, transportation, accumulation, and shipping. Each of these operational areas can be broken into transportation, accumulation, transfer, and control functions. Each of the previously discussed conveyor types fits one of these functions.

7.8.1 Conveying on an Incline (or Decline)

Pallets do not lend themselves to conveying on inclines or declines. Exceptions have been made with stable, well-stacked, and well-interlocked heavy carton loads. Some palletized bagged material can also be handled on slopes if the product is well stacked and interlocked. Keep in mind that any entrance to an incline, as well as the transition from the incline to a subsequent horizontal, causes an impact that affects the carton contents, load stability, and the pallet itself. Typically, drums are limited to grades of less than 10 degrees; all others are limited to 5 degrees if the loads can handle the transitions.

7.8.2 Conveyable Products

Pallets are far and away the most popular carrier for heavy unit loads. Pallets can be constructed of wood, fiber board, plastic, steel, or aluminum.

7.8.2.1 Conventional Wood Pallets

Wood pallets are the most frequently used pallet type. The different types of conventional wooden pallets are described in the following paragraphs, as are other commonly used pallets.

There are a variety of wood pallet types. Here, they are broken down by construction type.

- Stringer pallets. Stringer pallets are made up of a top layer of boards that are nailed to two or three stringers. Reusable pallets typically have a layer of boards on the bottom of the stringers. Four-way pallets, shown in Figure 7.12, are the most popular pallet design for grocery and dry-goods distribution. The stringer is notched to allow fork truck access from all four sides. A 40″ × 48″ pallet of this design is referred to as a GMA (Grocery Manufacturers Association) pallet. Figure 7.13 shows a two-way pallet. Notice that the stringers are solid so a fork truck can only pick it up from either end.

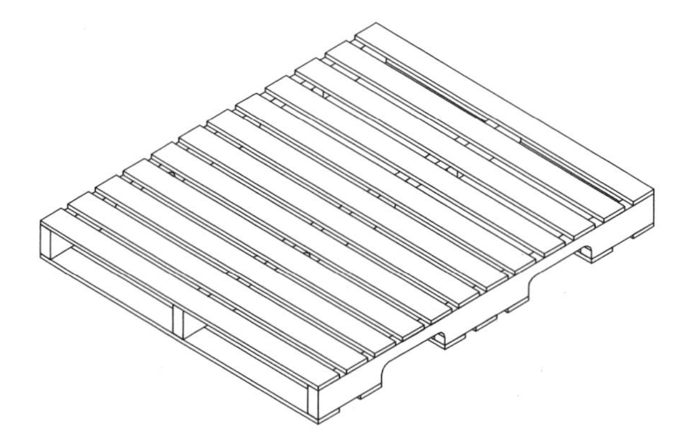

FIGURE 7.12 Four-way pallet.

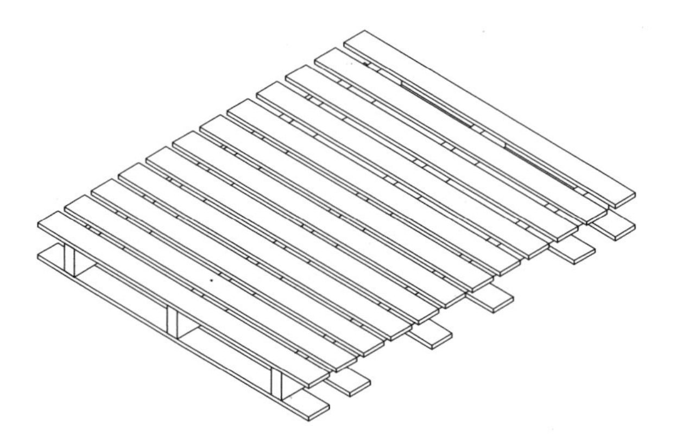

FIGURE 7.13 Two-way pallet.

- Block pallets. Block pallets typically have nine square blocks arranged with one in each corner, one in the center of each side, and one in the middle of the pallet (see Figure 7.14). This type of pallet has a plywood top and typically a plywood bottom. CHEP is a standard design of a 40″ × 48″
- Disposable pallets. Disposable pallets are usually a variation of one of the other standard pallet types. They are less expensive and are made with inferior materials. Disposable pallets are typically designed to handle the product through just one-use cycle and then are disposed of.

It is important to make sure that the bottom boards of pallets are parallel to the product flow on roller conveyors and perpendicular to the product flow on chain conveyors. Because all materials will flex as weight is put on them, it is important to ensure that pallets can properly support themselves. The bottom boards should not deflect more than ⅛ in. when on roller conveyors and less than ¼ in. when carried on chain conveyors.

When necessary, pallets with their bottom boards perpendicular to flow can be carried on roller conveyors as long as there are at least five rollers supporting the

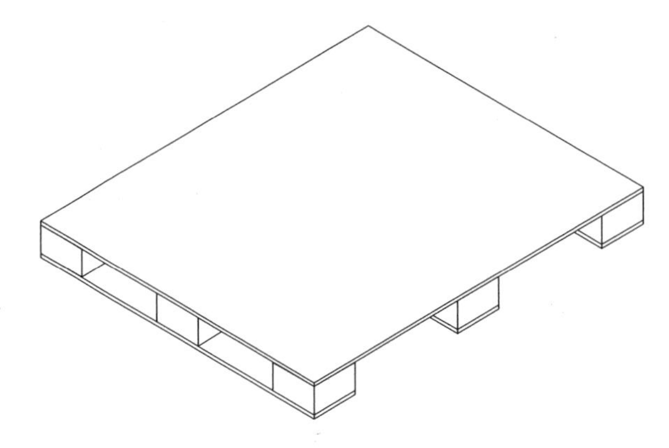

FIGURE 7.14 Block pallet.

pallet at all times. When carrying a pallet in this manner, it becomes even more important to make sure that only five rollers have the capacity to support and drive the load.

Wooden pallets offer a variety of advantages such as low per-unit cost and ability to be repaired. They are recyclable as well as biodegradable. The primary disadvantages are that they are not as durable as metal or plastic pallets, and as they are used and abused, broken boards and protruding nails can cause the pallet to get hung up on the conveyors.

Standard pallet sizes for various regions of the world are as follows:

North America	48 in. × 40 in. (1219 mm × 1016 mm), 42 in. × 42 in. (1067 mm × 1067 mm), and 44 in. × 56 in. (1117 mm × 1422 mm)
Europe	1200 mm × 800 mm, 1200 mm × 1000 mm, 1140 mm × 1140 mm
Asia	1100 mm × 1100 mm

When selecting a pallet size, it is important to not only consider the product to be handled, but also how the pallet will be used. If it will be used for shipping product, will the size fit in a trailer or export container? When dealing with export containers, it is best to determine a method that maximizes the use of the cubic space available.

Another concern with single-use or disposable wood pallets is that as the bottom boards get thinner, they begin to conform to the conveying surface. If the boards get thin enough, they will not be able to properly support the load.

Typical applications for wood pallets include distribution of groceries and dry goods, automotive hardware, and other durable goods.

There are three standard 40″ × 48″ pallets in North America. The GMA was mentioned earlier. The other two are CHEP and PECO. Both are companies that rent the pallets. CHEP (Commonwealth Handling Equipment Pool) is an Australian company with offices in the United States and the United Kingdom. PECO is in the United States and Canada. Similar in design both are block pallets; the distinguishing difference is CHEP paints their pallet blue and PECO's are red.

In Europe, there is EPAL (European Pallet Association) based out of Germany. Again, theirs are block-style pallets in the standard European dimensions. Their pallets are branded with the company name on one of the corner blocks.

7.8.2.2 Metal Pallets

Pallets made of steel or aluminum have a variety of advantages: They are very durable compared to wood pallets, are more sanitary because they do not absorb moisture or dirt, and can be recycled. Although these attributes are commendable, metal pallets have disadvantages as well (see Figure 7.15). Metal pallets are more expensive, they are generally not repairable, and they generally do not convey well if damaged or deformed.

Typical applications for metal pallets include heat-treating furnaces, in-process manufacturing operations, chemical and food processing applications, automotive manufacturing, military uses, and closed-loop systems that automatically recycle pallets within the system.

7.8.2.3 Plastic Pallets

Plastic pallets come in the widest variety of shapes. Some are shaped like typical wood pallets, but others are molded with legs and still others are vacuum molded for a specific application (see Figure 7.16).

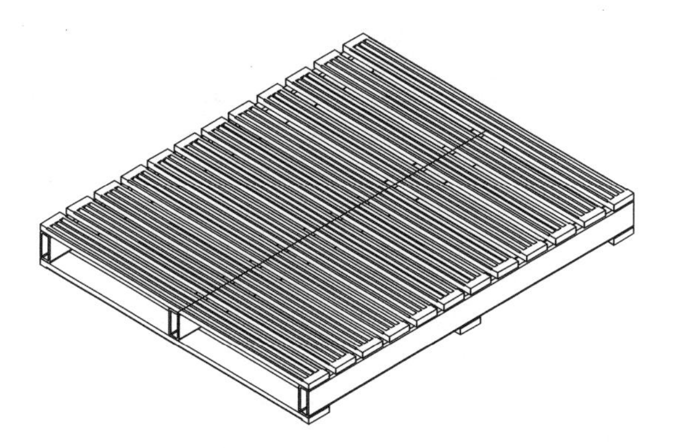

FIGURE 7.15 Metal pallet.

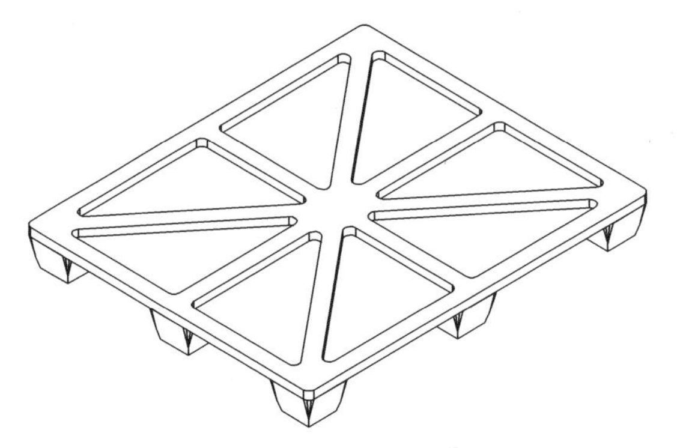

FIGURE 7.16 Plastic pallet.

Plastic pallets have many of the same advantages as metal pallets: They are very durable, more sanitary than wood, and they can be designed to be nestable for reduced storage space. They conform to the conveying surface and can be custom designed to carry specific products like drums and car seats. The drawbacks of plastic pallets are that they are costly, they are not always recyclable, they are not repairable, and they can deflect more than wood or steel pallets.

7.8.2.4 Fiber Boards

Fiber board pallets, like plastic pallets, come in a wide variety of shapes. Some might be molded with legs, and others can be simple, flat, slave pallets.

Fiber board pallets have some advantages, such as being lightweight, recyclable, and biodegradable. They can also be designed to be nestable for reduced storage space. The drawbacks of fiber board pallets are that they are not suitable for outdoor use, they can break down when wet, and often they are not conveyable.

7.8.2.5 Slave Pallets and Slip Sheets

Slave pallets are typically ¾-in. to 1-in. thick. With no openings, it is very difficult for forklifts to handle them. Typically, a clamp truck or other specialty truck is required to handle loads on a slave pallet.

Slip sheets are basically reinforced sheets of cardboard. These are typically used under unitized loads of large bags of dog food or paper products. Like slave pallets, slip-sheeted loads have to be handled by clamp trucks. The primary drawback to conveying loads on a slip sheet is that because the slip sheet is flexible and conforms to the roller, it tends to creep out from under the load. Slip-sheeted loads cannot be handled on a chain conveyor without some form of support between the chains. The alternative to a chain transfer is to use a very large belt turn. A CDLR roller turn will not work well because the rollers are too far apart.

7.8.2.6 Containers

With the advance of kitting concepts and just-in-time systems, larger containers have become more prevalent. These containers are typically 40 × 48 in. and of varying depth. Gaylords, large cardboard boxes attached to a wood pallet, are very popular. They are used to carry bulk items such as small parts. Containers can also be made in the form of large wire baskets or assembled from large, injection-molded plastic components. In all cases, the ability to collapse the empty containers to return them back to the source is a great cost saving.

7.8.2.7 Drums

Besides being conveyed on pallets, drums can be conveyed by themselves. Steel, plastic, and fiber board drums are popular containers not just for liquids but also for bulk dry materials. The drums come in a variety of sizes that range from 55 gallons down to 30 gallons. The one thing that they have in common is a chimed bottom, where the outside rim of the container extends below the rest of the container. The container is supported by the rim. This presents a challenge on multistrand chain conveyors. Due to possible leakage, drums should not be carried on lineshaft or padded chain-driven live roller conveyors.

The effective conveyor width of roller conveyors should be 1½ to 2 in. wider than the drum diameter, which prevents the edge of the drum from extending beyond the end of the roller. Due to the greater supporting surface, larger drums can be allowed to extend beyond the edge of the rollers, but it is not recommended. Drums can be carried on multistrand chain conveyors. Keep in mind that the outer edges of chimed product must be supported at transition points.

Verify the condition of drums to be conveyed: Are they new or reconditioned? The rims must not be bent or dented, as they might not convey properly. The contents must be considered when selecting the conveyor because there will almost always be spillage or leakage. Never convey drums on their sides, as the drum surface is very vulnerable to denting.

One conveyor system I did for Kodak™ was conveying drum of photographic chemicals. For this reason, the entire CDLR conveyor was constructed of stainless steel, the frames, guards, rollers, and roller chains.

7.8.3 System Development

The primary key in developing an HULH system is to ensure the product is properly supported at all times. This typically translates into using CDLRs when the bottom boards are parallel with the direction of product flow and multistrand chain conveyors or CDLR with very close roller centers when they are perpendicular to the product flow.

Because CDLR is less expensive than multistrand chain conveyor, it is most cost-effective to design the system so that the pallet is oriented to allow the vast majority of the conveyor to be CDLR. If the product is hand-stacked pallets of boxes, it is simply a matter of the pallets' original orientation in the stacking workstation. On the other hand, if the product will be coming out of a palletizer, then it is up to the palletizer manufacturer.

Typically, palletizers are designed to produce stacked pallets that are oriented with the bottom boards perpendicular to the exit. This means that a multistrand chain conveyor would be used at the exit. This is acceptable if we can make a right-angle transfer shortly after exiting the palletizer. This is frequently the case when multiple palletizers are placed side by side and they all discharge toward a common takeaway line. In Figure 7.17, two bulk palletizers and a carton palletizer all exit onto a common CDLR conveyor that feeds a stretch wrapper.

7.8.3.1 Pallet Loading

Pallets can be loaded manually or mechanically. The most common pallet load is a stack of cartons. These cartons can be stacked in a variety of patterns, depending on the size of the cartons. Figure 7.18 shows various stacking patterns. Column or block stacking provides the least stable load, whereas the more interlocked the pattern is, the more stable the load is. As mentioned previously, a carton palletizer is frequently employed to create these patterns automatically.

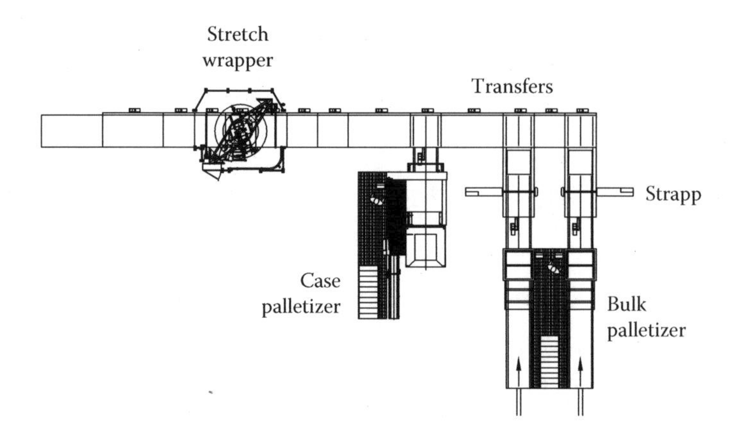

FIGURE 7.17 Heavy unit load conveyor system example.

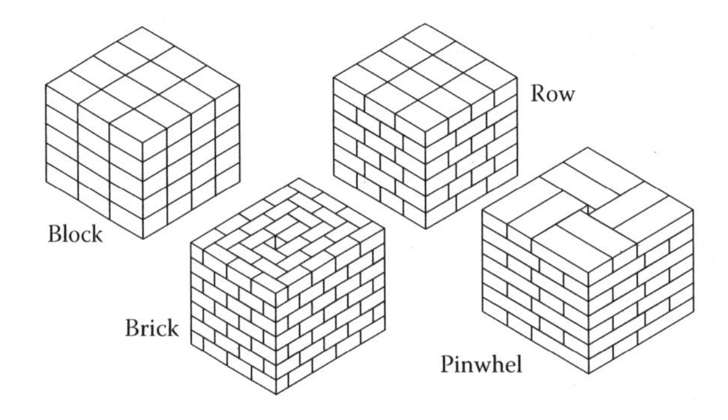

FIGURE 7.18 Common palletized carton patterns.

To increase the stability of a load, there are several methods:

- Spiral stretch wrapping
- Plastic or metal strapping
- Shrink wrapping

There are other choices available, but these are by far the most popular.

When multiple products are stacked on a pallet, it is always considered best practice to keep all products within the footprint of the pallet. Due to the possibility of the product getting snagged on something as it is being conveyed, any overhang should be kept to a maximum of 50 mm (2 in.). It is also advisable to have the heaviest product on the bottom of the stack with the lighter or fragile product on top.

Table 7.4 and Figure 7.19 describe several types of unstable loads that require attention.

When palletizing by hand, it is important to keep in mind how the product will stack. It's really like a 3D game of Tetris®. Some companies employ what is referred to as bang boards. These are a pair of walls at a right angle that give the person palletizing a sturdy right-angle corner to stack against to help ensure a neatly stacked load.

TABLE 7.4
Unstable Load Types

Load Type	Problems
Leaning (shingled)	This type of load is of special concern with block stacks. This can get caught easily and toppled by careless handling of surrounding loads.
Opening up	This is common with manually palletized loads. This can get caught easily and toppled by careless handling of surrounding loads.
Wobbly load	This is common with product that is not firm and flat. This happens frequently with bagged product that is not properly flattened prior to stacking.
Bulging load	This is caused by vertical loading. The product is not strong enough to support that many layers.

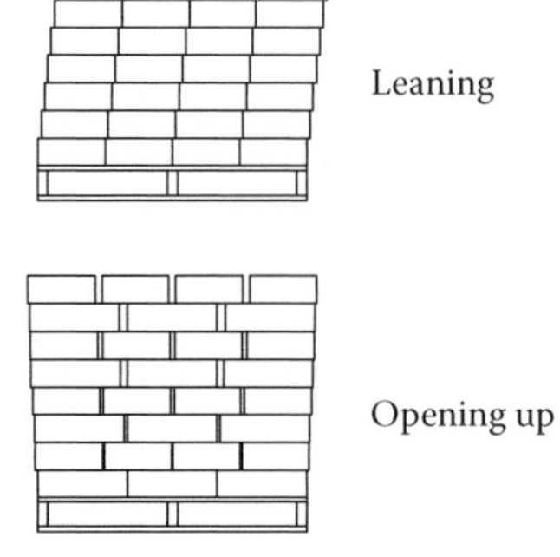

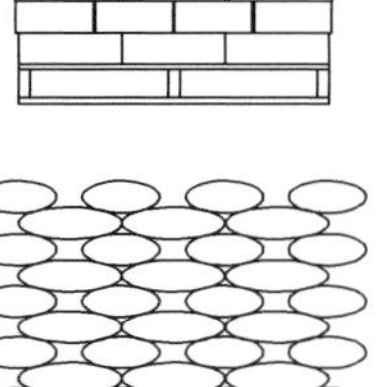

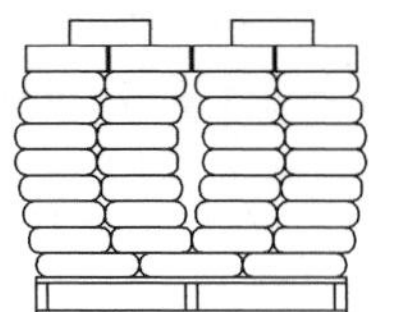

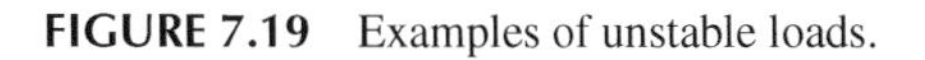
FIGURE 7.19 Examples of unstable loads.

7.9 QUESTIONS

1. If the bottom boards of a pallet are perpendicular to the direction of flow, what is the simplest conveyor of choice?
2. In the previous question, what would be one of the criteria for a CDLR conveyor?

3. What is a common way to stabilize an unsteady load on a pallet?
   a. Strapping the load
   b. Stretch wrapping the load
   c. Shrink wrapping the load
   d. All of the above
4. If CDLR with 2 ½″ rollers on 3″ centers would work regardless of pallet orientation, why not use it everywhere in a system?
5. Why is a CDLR curve not recommended for handling slip-sheeted loads?
   a. It cost more than a belt turn.
   b. A chain transfer will do the job well.
   c. The rollers are spaced too far apart.
   d. Trick question; a CDLR curve is very good for slip-sheeted loads.
6. The GMA (Grocery Manufacturers Association) standard is the 48″ x 40″ four-way pallet is used in the US and EPAL standard is the 1200 mm × 800 mm pallet used in Europe. How would you ensure that a conveyor system can handle both if it had to?
7. What type of conveyor would you select to accumulate 2,500-pound pallets of beverages?
   a. Padded chain-driven live roller
   b. Lineshaft conveyor
   c. Chain-driven live roller
   d. Belt-driven live roller
8. What type of conveyor would select to accumulate 1,000-pound pallets of snack chips?
   a. Padded chain-driven live roller
   b. Lineshaft conveyor
   c. Chain-driven live roller
   d. Belt-driven live roller
9. What are some aspects of the product on that pallet that must be considered when selecting a type of accumulating conveyor?
10. What are the apparent trade-offs between CDLR curves and a CDLR conveyor on a turntable?

8 Overhead Chain

Overhead chain conveyors offer advantages not found in traditional roller or belt conveyors. Overhead chain conveyors leave valuable floor space free. Also, many of the products that can be handled by an overhead chain system would be considered non-conveyable on a traditional conveyor.

The three types of overhead chain conveyors are free, power, and power and free (P&F). The names refer to how the carriers on the conveyors are moved.

All of these conveyor types have basic system design requirements that must be dealt with up front. The primary design requirement or consideration is carrier design: how is the product going to be supported by the conveyor? As the carrier follows the overhead track, the product must always stay on or in the carrier so the carrier must balance the load on the conveyor. In some applications, the carrier must be able to orient the product, such as rotating a product in a paint booth or an interface to another piece of equipment.

The second design requirement is carrier and product spacing. For straight sections of a conveyor, spacing is only a minor concern. However, once curves or inclines are introduced spacing can become a primary concern. Carrier design plays into the spacing as well (see Figures 8.1 through 8.4).

One of the most prolific uses of P&F systems, both overhead and inverted, occurs in automotive assembly plants.

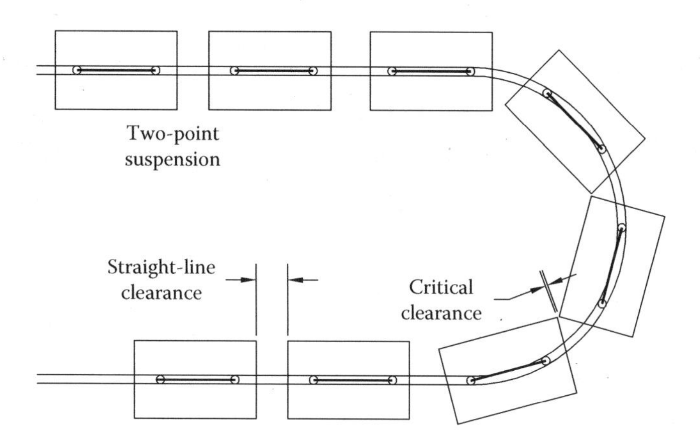

FIGURE 8.1 Product clearance in a two-point suspension system.

DOI: 10.1201/9781003376613-8

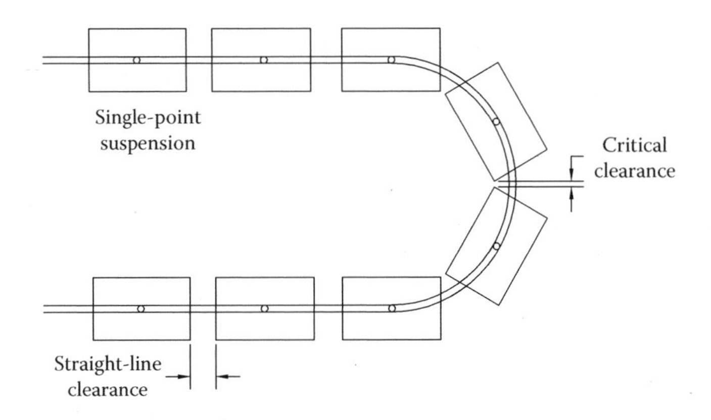

FIGURE 8.2 Product clearance in a single-point suspension system.

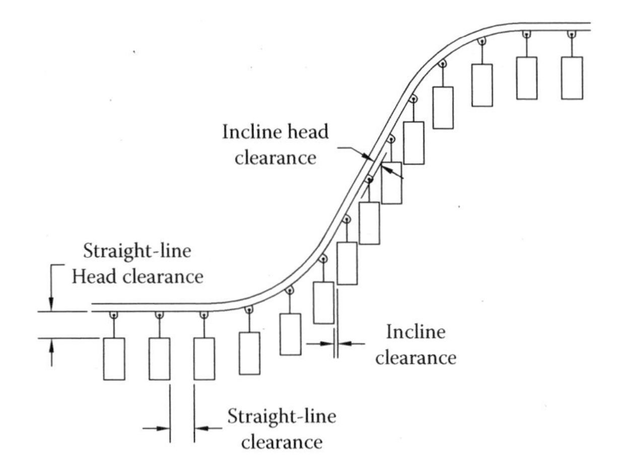

FIGURE 8.3 Product clearance in an incline/decline single-point suspension system.

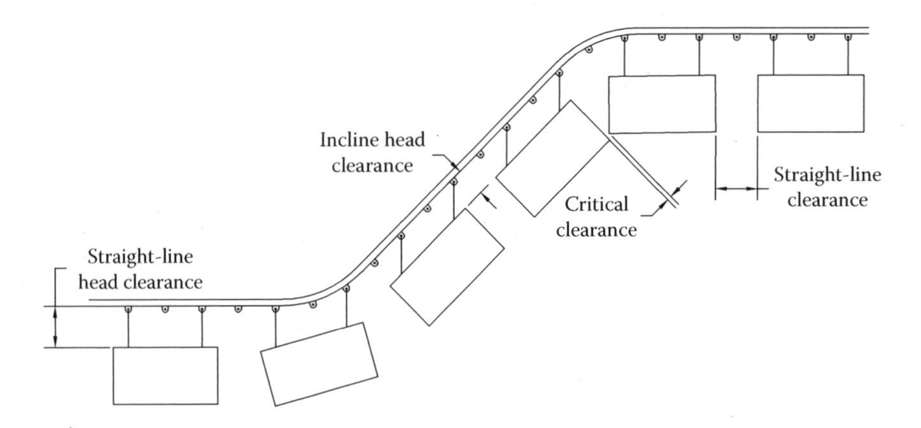

FIGURE 8.4 Product clearance in an incline/decline two-point suspension system.

8.1 FREE SYSTEMS

The first member of the overhead conveyor family is the Free overhead conveyor. Although it is the least-expensive type of system, the "Free" title refers to the fact that the carriers are free to be moved along the rail individually. There is no chain connecting the carriers. In a free system, the product is moved either manually or by gravity. Gravity sections of free conveyor should be used very judiciously because heavy loads or many light loads can get out of hand.

Free systems are suitable for low-production situations, small installations, and manual work or storage areas of a P&F system. Typical applications include portable tool supports, small batching-type paint systems, assembly, inspection and testing work cells, and product storage.

It is important to keep loads down to a weight that can be moved manually. For obvious reasons, free systems should be used in areas that are accessible to workers.

There are five basic components to a free system. The first is the straight tracks. There are two general types of tracks for overhead chain systems: open and enclosed. The open track is typically a structural steel beam (e.g., an I-beam) that uses a four-wheeled trolley to hang below the beam, as shown in Figure 8.5. The enclosed track uses round or square tubing with a slot cut in the bottom to allow the carrier to ride inside. Both have their advantages and disadvantages, as illustrated in Table 8.1.

The second and third components are the horizontal and vertical curved tracks. Horizontal curves maintain the same elevation but allow the carriers to turn corners. Vertical curves allow the carrier to change from a horizontal path to an incline or decline path. Both are rolled versions of the straight track, and both are available in various radii. As mentioned earlier, not only the layout but also the product and carrier help determine the radius of the curves that can be used.

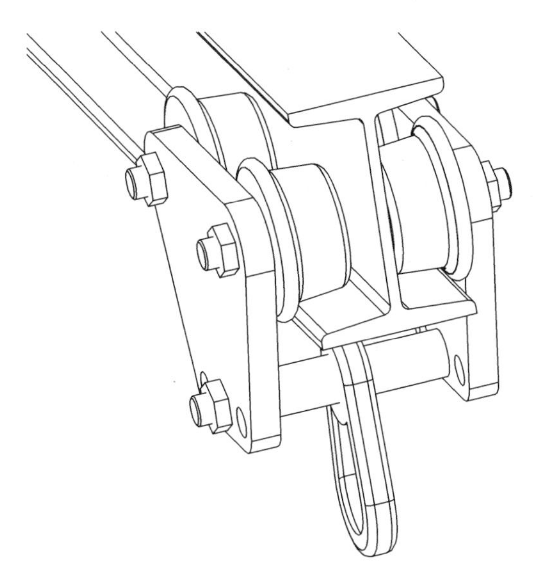

FIGURE 8.5 Example of an open track-free carrier.

TABLE 8.1
Track Type Advantages and Disadvantages

	Advantages	Disadvantages
Open track	• The rail is readily available structural steel • Higher weight capacity • Only relatively large radius turns are available	• Chain and carriers are open to environment
Enclosed track	• Helps prevent accidental contact with the chain • Keeps foreign debris out of the chain • Tighter radius turns are available	• Track is more expensive

The trolley is the fourth element in a free system. This is the component that rides in or on the track and supports the product. There are a wide variety of carriers; each fits a specific type of track. The load to be carried will help determine the trolley type and the track to be used. Trolleys in a manually powered free system typically have four wheels to minimize swing. How the carrier will attach to the trolley is also a consideration. Does it need to be able to rotate or must it stay rigid?

The next component of a free system is the carrier. As already mentioned, carrier design is a primary requirement. An important factor is how the product is going to be supported. The product must always stay on or in the carrier so the carrier must balance the load beneath the track. The simplest carrier is just a hook hanging from the trolley on which the product is hung. In other cases, the carrier may hold multiple parts on shelves or in baskets (see Figure 8.6). If the carrier must interface with other machinery, stability and repeatability is crucial.

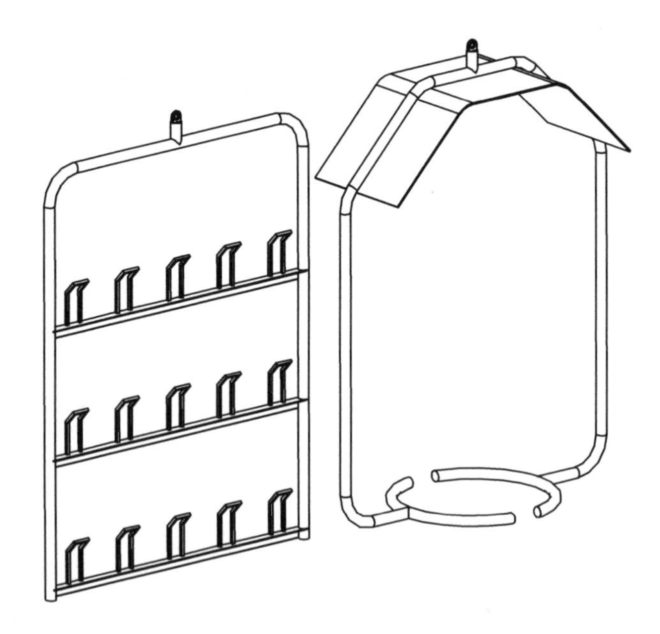

FIGURE 8.6 Sample carriers.

Two additional components that are used primarily with enclosed track systems are diverts and merges, also referred to as switches and frogs. All can also be referred collectively as switches.

Divert-type switches can be controlled manually or by an air cylinder or electric solenoid. Diverts are used to direct trolleys from the main line to a separate line such as a workstation. Merges are not powered for enclosed tracks. The trolley simply pushes the center pivot over to provide the trolley with the necessary support. For the open track, a merge is simply a divert installed in the opposite orientation.

8.2 POWER SYSTEMS

In a power system, the carriers are linked together by a chain. The chain is driven to make the carriers move. Power systems are well suited for medium- to high-production areas with continuous product flow. Typical applications include oven and freezer lines, paint lines, or any process where the product does not have to stop. Stopping one product means stopping the entire conveyor. This means that the product has to be loaded and unloaded from a moving carrier.

A power system can use the same components as a free system, with the addition of three components: the chain, the drive, and the take-up. The other difference is that the trolleys of an open track system have two wheels rather than the four wheels mentioned earlier.

In an open track system, the chain is typically a heavy cast link chain that connects the trolleys. By contrast, in an enclosed track system the trolley becomes an integral part of the chain. As illustrated in Figure 8.7, unlike the heavy cast chain that connects the trolleys of an open track system, this chain has wheels mounted both vertically to support the load and horizontally to minimize friction through the curves. The wheels can be steel or have a nylon face to minimize noise and track wear. The trade-off is that the nylon wheels have a lower load rating.

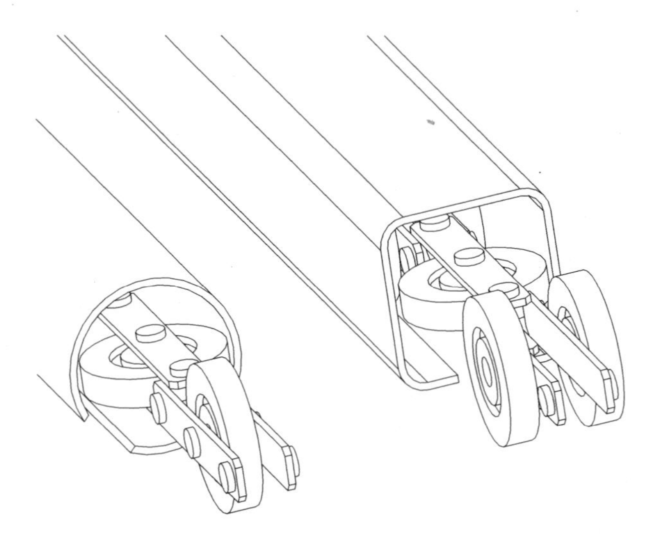

FIGURE 8.7 Examples of closed-track-powered chain.

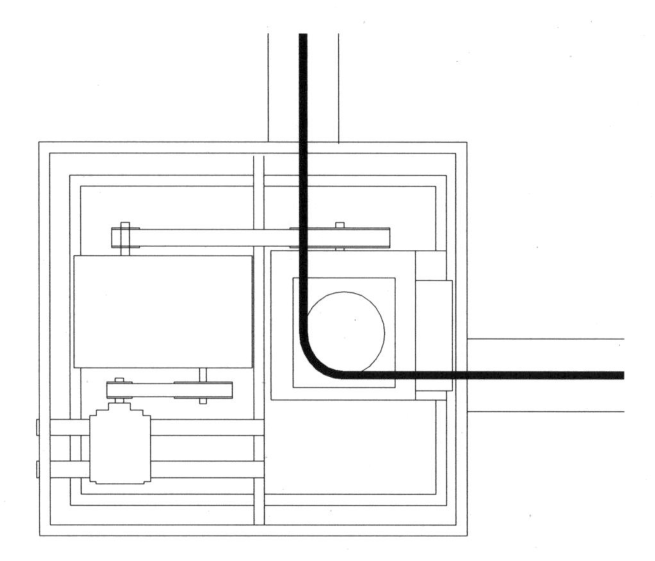

FIGURE 8.8 Ninety-degree sprocket drive.

The drive is a component added along the length of the system to pull the chain. Drives typically come in one of two forms. The first is a sprocket drive (see Figure 8.8). The sprocket drive is designed to guide the chain so that it wraps around a custom sprocket. The sprocket drive can be designed to mount in a straight section, a 90-degree corner, or a 180-degree corner. The straight-section version can mount anywhere along any straight section but offers the least drive capacity. The 180-degree corner version offers the highest drive capacity but is the most limited in location. The sprocket drive is used primarily with open track systems, although there are some manufacturers that offer it for enclosed track systems.

The second drive type is the caterpillar drive. This type of drive mounts along any straight section. The caterpillar drive has a short, driven, specialty chain that engages the conveyor chain, thus transferring the drive power (see Figure 8.9). In open track systems, the chain is engaged from the side and in enclosed track systems the chain is typically engaged from above. Regardless of drive type, if the chain pull of the

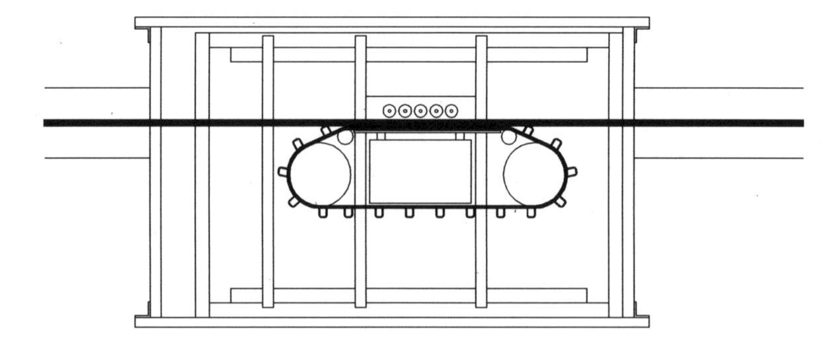

FIGURE 8.9 Caterpillar drive.

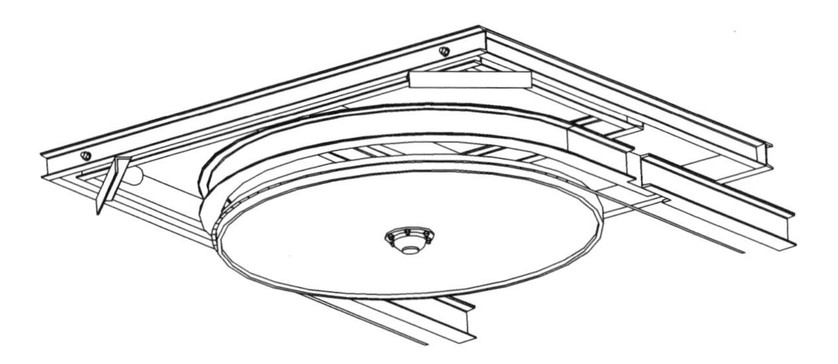

FIGURE 8.10 Take-up curve.

system is high enough, multiple drives can be used in a single system. With multiple drives, they must be driving the chain at the same speed.

The last new component is the take-up. As with any chain in any application, the chain stretches and the system must have a means to take up the slack caused by that stretching. In addition to compensating for chain stretch, the take-up also provides compensation for chain growth or shrinkage due to temperature changes. The take-up is a 180-degree curve that is mounted in a rigid frame that allows it to move freely (see Figure 8.10). A take-up typically uses either a spring or a counterweight to keep the chain tensioned properly. The take-up is positioned so that the chain goes through it shortly after the drive. The length of take-up travel that is required depends on the overall length of the system and the temperature variation that the system might see. Spring take-ups are used for typical take-up requirements. Counterweighted take-ups are frequently used where more movement is required. This is the case in a system that might go through a furnace, where the chain will grow considerably due to the heat and shrink back when the furnaces are shut off.

Note that power systems do not use switches because in a power system the chain is one continuous loop.

A specialized type of a power system is referred to as GOH (garment on hanger). These are popular not only at your local dry cleaner but also in distribution centers for apparel. These are light-duty overhead chain conveyors.

8.3 POWER AND FREE SYSTEMS

P&F systems combine the features of both free and power systems. A P&F system actually uses two tracks rather than just one. One track supports the trolleys and the other supports the chain. The chain is not permanently attached to the trolleys as it is in a power-only system. This separation of the trolleys and the chain opens up a world of application possibilities.

The primary difference between a power system and P&F system is that there are two tracks rather than one (see Figure 8.11). This allows the trolleys to ride in one track separate from the chain. Protrusions from the chain, or dogs, engage the trolleys and pull them along.

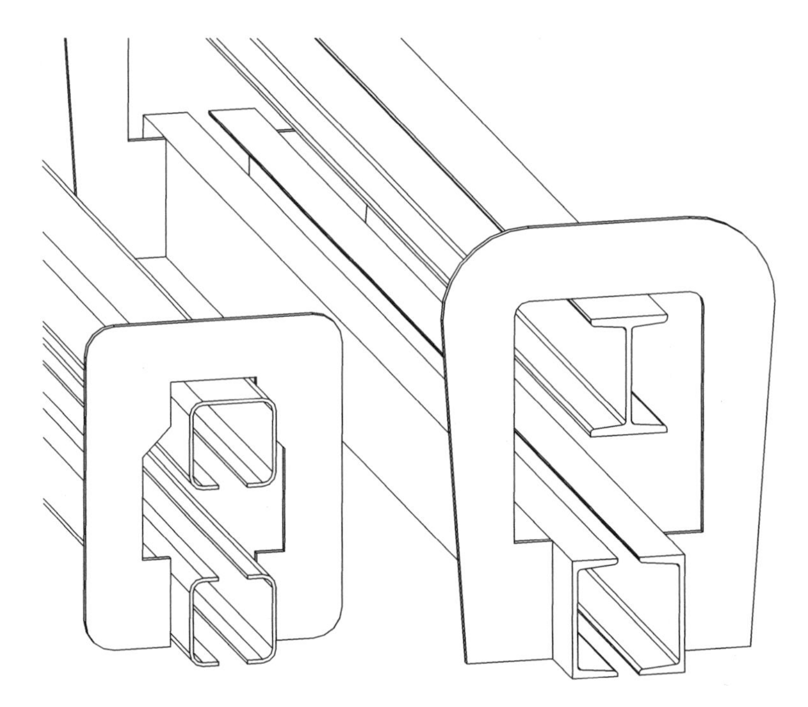

FIGURE 8.11 Example of a closed track and an open P&F track.

Some systems are designed so the two tracks are side by side and others are designed so the track for the chain is above the trolley track. Most manufacturers offer the over and under version, making the side-by-side version far less popular.

Just as a power system added components to the free system, so too does the P&F system add components to the free and power systems. The first, as we have already discussed, is a second track. The second component is a stop, which is used to disengage the dogs from the trolleys. When the dogs disengage the trolleys, they stop moving under power and are free to move as in a free system.

To accumulate product, the system typically uses a stop to halt the movement of the first carrier. Then, due to the mechanical design of the chain dog and trolley, the next carrier comes along and the back of the previous trolley disengages the dog from the approaching trolley, thus making it stop (see Figure 8.12). This will continue with each successive carrier. Once the stop that initiated the accumulation is released, the trolleys are reengaged by the next dog that comes by. When the first carrier has moved on, the next one will be reengaged by the next dog, and so on.

For larger or heavier products, carriers can be hung from multiple trolleys as shown in Figure 8.13. Notice that the trolley feature that disengages the dog from the next carrier for accumulation is moved to the rear trolley. The front trolley still has the catch point so that the trolleys are pulled, not pushed. Three or four trolleys can be used to carry even heavier loads. Intermediate trolleys do not have either feature on top. The key is that the loads can navigate the horizontal and vertical curves.

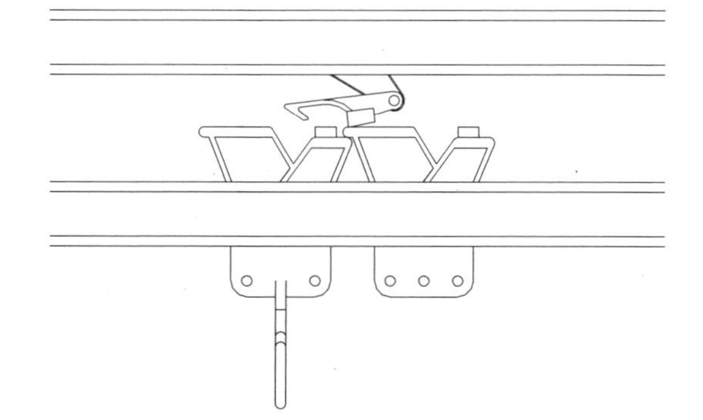

FIGURE 8.12 Example of dog engagement.

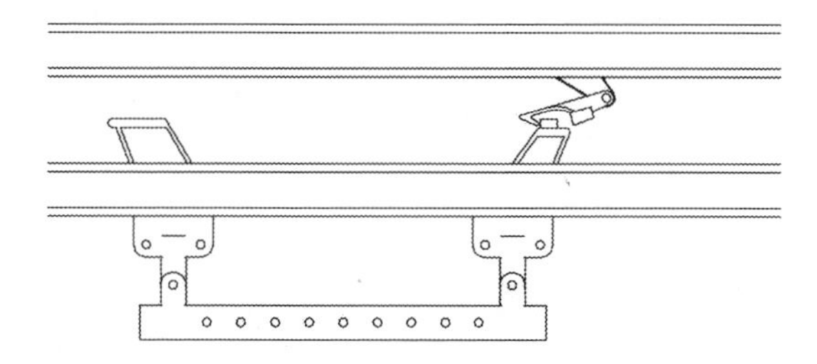

FIGURE 8.13 Two-point carrier arrangement.

Switches are used in the track on which the trolleys ride. Diverts provide an exit point from the main line for trolleys to enter a workstation or storage bank or even transfer to a different powered chain. Again, these switches can be controlled manually or by an air cylinder or electric solenoid. The manual switch is frequently used so operators can divert carriers into their workstations. The electric and pneumatic switches are used where the control system has direct control of the switch.

Merges are not powered but do require a stop upstream of the merge to prevent trolleys from entering the main line before there is an opening for them on the chain. This prevents jams and collisions.

8.4 ACCESSORIES

A wide variety of accessories are available to aid the functionality of power and free systems. Stops and switches have already been discussed.

Closely related to the stop is the escapement. The concept of an escapement is that it is an alternating stop. It releases one carrier and only one at a time. When the escapement is at rest, a carrier travels to the end of the track and is stopped by the escapement and others can accumulate behind it. When the escapement is engaged the first carrier is released, but at the same time the second carrier is held back so it is not released. Escapements are frequently applied at the discharge end of an accumulation line or work cell directly prior to a merge.

Another important accessory is the anti-rollback device. This is a simple device used on free or P&F inclines to prevent the trolleys from running back down the incline.

Inspection stations are sections of track that can be opened to allow inspection of the chain.

An installation gate, normally positioned at the exit side of a drive, is used to install or remove sections of chain. At least one is required in every system.

Another accessory is the expansion joint. These are used in a system wherever the system crosses the expansion joint of an oven. This allows the track to expand and contract with the oven structure as it heats and cools. The centerline of the expansion joint should be located on the centerline of the oven's expansion joint.

Automatic lubricators are important accessories used to keep the chain lubricated in power and P&F systems.

The final accessory that we will discuss is the carrier clamp. In a system that must interface with other equipment such as other conveyors, robots, or some other outside operation, the carrier can be clamped into a fixture to ensure proper, repeatable locating. A carrier clamp is used when the carrier is in its free condition, never when it is being moved under power.

8.5 INVERTED POWER AND FREE

Inverted power and free (IP&F) conveyor systems provide all the benefits of overhead P&F systems along with several distinct advantages.

IP&F systems provide complete accessibility to the product, which makes them ideal for use with automation for assembly operations or painting. No bulky carriers or other hanging devices are required that could interfere with operators or robotic operations. Systems tend to be cleaner because no conveyor debris falls on the product. Systems are usually floor mounted, thus eliminating costly ceiling reinforcement.

IP&F conveyors can be mounted in pits or trenches and equipped with floor covers to allow traffic across aisles. Many buildings today are leased so cutting into the floor is not acceptable to the landlord/property owner, so this needs to be researched thoroughly.

The primary difference is that the carriers are equipped with wheels, casters, or some means of keeping the load stable. The carriers for IP&F are typically designed to be wide and more stable than those used on an overhead P&F system because with a narrow center support, additional stability must be added.

IP&F systems can be designed to operate as an in-floor tow line, pulling carts around the system. This allows the carriers to be easily removed from the system, typically through the use of a drop pin or some other quick release. The carrier can be rolled away from the conveyor and then later returned and reconnected to an available trolley.

8.6 APPLICATION

Unlike most conveyors discussed in this book, overhead and inverted systems focus more attention on the carrier than the product itself. Once the carrier is determined, the system must be designed to handle the carrier. Determine the optimum number of parts that will be on a carrier, making sure that the weight does not exceed the conveyor capacity. The carrier should permit easy loading and unloading. As mentioned at the beginning of the chapter, the load must be balanced. Load bars can be used to connect two trolleys with a carrier hung from them, thereby sharing the load. For longer loads, three or even four trolleys can be connected. The length of the load bars must take into consideration the radius of horizontal and vertical curves. Carrier design and track choice are frequently worked on simultaneously because one has a direct effect on the other.

Keep load clearances in mind when determining the radius of horizontal curve and the angle of incline or decline for elevation changes. This information might require reconfiguring certain parts of the system based on these restrictions.

From here on, the remaining steps are fairly straightforward:

- Determine guarding requirements to protect personnel.
- Determine the system rate, in carriers per minute.
- Determine carrier spacing.
- Determine maximum conveyor speed.
- Determine quantity of carriers based on conveyor length.
- Determine the number of loaded and unloaded carriers and thus total live load based on operations.
- Determine load due to elevation changes based on the layout.
- Determine chain pull based on data already gathered.
- Determine number of and location of drives required based on chain pull.

This is a brief synopsis of how a system is designed. This process holds true for both overhead and inverted systems. No calculations are provided here because every manufacturer has their own calculations for horsepower based on their own chain and trolley designs.

Some overhead systems have been designed so that the carrier actually rides on casters on the floor. This allows the casters to support the weight of the product while the overhead chain only has to pull the carrier along. This greatly reduces the horsepower required to power the system. Another advantage is that in a P&F system, a switch can be used to release a carrier from the system so it can be rolled around freely and then reintroduced through a merge point. The primary drawback is that it takes up more floor space and virtually eliminates the possibility to go over obstacles unless the carrier is empty prior to being lifted off the floor.

8.7 QUESTIONS

1. When using a P&F through a paint booth, what would be some of the considerations when selecting a carrier?
2. How can you carry a product on an overhead chain conveyor that is too heavy for an individual carrier?
3. If you had to unload a carrier using a robot, why couldn't you simply use an overhead chain conveyor?

4. When installing an IP&F as an in-floor toe line, can you think of any issues that need to be communicated between the conveyor company and the flooring contractor?
5. True or False: a combination of carrier spacing and product size are two determining factors on the angle of incline for an overhead chain conveyor.
6. Which of these is not part of an overhead chain system design?
   a. Determine the system rate in carriers per minute
   b. Determine the lubrication requirements for the drive
   c. Determine the maximum conveyor speed
   d. Determine the number of loaded and unloaded carriers and thus the total live load based on operations
7. In a furniture factory, which style of conveyor track would be better: open or enclosed? Why?
8. In planning an overhead chain conveyor for a powder coating system, after the powder is applied the parts must go through a 400° F furnace to melt the powder into a cohesive coating. What type of take-up should be specified?
9. An overhead chain conveyor is being used to deliver empty cartons to a series of pack stations. What specific data is not required to design a carrier?
   a. Minimum and maximum box sizes.
   b. Rate at which the boxes will be consumed.
   c. How are the boxes going to be erected?
   d. How high above the floor will the conveyor chain be mounted?
10. True or False: merges and/or diverts can easily be added to a basic powered overhead chain system.

9 Miscellaneous

The conveyors listed here do not really fit into any of the other categories in this book. They are important and fill specific material handling needs. I have avoided conveyors that are specific to a particular industry and have instead included conveyor types that are applicable to multiple industries, which are of interest to the broadest audience.

9.1 CHUTES

Chutes are used as an inexpensive way to move products without electricity. Chutes use gravity as their driving force. That dictates that the surface of the chute has to have a low coefficient of friction to minimize drag on the product. There are many different materials used for chutes and slides such as plastic laminates (Formica), steel, stainless steel, and various plastics. By far the most popular is steel. Each has its advantages and disadvantages, as detailed in Table 9.1.

Chutes can be straight or curved. Straight chutes are the most popular and also the easiest to manufacture. They can be as short as a few centimeters or several meters long. Straight chutes are also the least expensive.

When a large change in elevation is required in a short distance, spiral chutes are an excellent solution. They can be made of steel or fiberglass and are typically curved around and mounted to a central pole. Spiral chutes offer several advantages over typical straight chutes and have some drawbacks. Spiral chutes can handle input at several points along their length. Properly designed spiral chutes control the speed of the product's descent due to centrifugal force pushing the product outward to a longer, shallower path. This same force pushes it against the outside rail, creating drag. Unlike steel, the effectiveness of fiberglass chutes is not seriously impacted by changes in environment such as humidity. The gel coating used in a fiberglass chute is typically flammable, although some manufacturers offer flame-retardant grades of gel coating for their fiberglass chutes. Steel chutes, by their very nature, are not flammable.

Environmental conditions where the chute will be installed are an important factor. When chutes are located in non-climate-controlled facilities or near dock doors, the product's travel characteristics can change dramatically in a short period of time. When humidity increases, corrugated cardboard becomes more pliable and will not move as readily. Another aspect of increased humidity or quick air temperature increases is that chutes can "sweat" or collect condensation on their surface. This can slow most products significantly. Any testing should incorporate such conditions. For steel chutes, this condensation can cause rust to form on the chute when it is not in use. The rust can transfer to the product leaving marks on it. The rust can cause permanent rough areas on the chute surface if left unattended long enough.

DOI: 10.1201/9781003376613-9

TABLE 9.1
Chute Material Comparison

	Advantages	Disadvantages
Unfinished mild steel	• Provides a low coefficient of friction	• Will oxidize in areas where product is not traveling
Painted mild steel	• Will not oxidize	• Requires additional treatment to provide low coefficient of friction
Stainless steel	• Provides a low coefficient of friction • No oxidation	• More costly than mild steel
Fiberglass	• Provides a low coefficient of friction	• More costly than steel when used for straight chutes
Plastic laminates	• Provides a low coefficient of friction	• Poor wear characteristics
UHMW	• Provides a low coefficient of friction	• Costly

Several aspects must be considered when applying chutes. Chutes, for obvious reasons, need to be steep enough to keep the product moving. Less obvious and equally important is that the chute is also steep enough to ensure that product will restart once it has been stopped in the chute.

Some of the rules change for bulk and unit loads. Typically, chutes being used for bulk products are tapered to funnel the product into a more contained flow. This helps reduce dusting and minimizes the size of the potential area where product will fall. With bulk materials coming off of a belt conveyor, it is important to consider the product trajectory coming off of the head pulley and provide hardened steel liners in and around the impact area. The liners should be replaceable.

For unit loads, on the other hand, funneling in a chute has to be done very carefully. Due to the large frictional area associated with unit loads such as cardboard boxes, the added drag from a guide rail or chute side might be enough to stall the product in the chute. If you make the chute steep enough to overcome this additional drag, it could be too steep for product that does not contact that side rail. Typically, a total included angle of 15 degrees is the recommended maximum for funneling chutes.

Typically, straight chutes for unit handling can be designed to keep the product properly oriented as it navigates the chute. Spiral chutes, on the other hand, do not offer the size flexibility of straight chutes and therefore the product orientation is seldom consistent. This factor must be heeded to prevent jams at the discharge of the chute. In some applications, adding a long-tapered guide rail at the discharge of the chute to help reorient the product without stalling it has proven effective.

When a chute is going to discharge onto another conveyor, it is always a good idea for the product to be moving in the direction of the conveyor if at all possible. With bulk conveyors, this greatly reduces conveyor impact, and on unit handling conveyors, it reduces jams. Also, when dealing with unit handling, it is best to have a “swoop” on the end of the chute that prevents the product from stubbing into the rollers.

Frequently a short section of gravity wheel or roller conveyor is used after a chute to help slow down the product. An alternative to control product speed in a chute is the addition of baffles or bumpers. Baffles can be as simple as vinyl strip curtain hung down into the path of the product to create additional drag. Bumpers can also be quite simple, such as a section of rough-top conveyor belting mounted to the side of the chute to create more drag in that area. This is particularly effective with spiral chutes, where larger, faster-moving product moves to the outside periphery of the chute. The belt can create the necessary drag to slow the product down and send it back toward the center of the chute's path.

With vertical bulk material chutes alternating baffles can be added to slow the product down. This is especially important with long drops.

9.2 VIBRATORY CONVEYORS

Vibratory conveyors are unique in that, for all intents and purposes, the conveyor components remain stationary and only the product being conveyed is moving. Using controlled vibration, product can be moved very reliably in a specific direction. The idea is that the forward motion is slightly upward and slow and the return motion is downward and fast. This moves the product forward and then leaves it almost hanging in the air on the return stroke. When the product falls back to the surface of the conveyor, it is ahead of where it was before. The motion with each stroke is very small and compressed together, thus creating the vibration. This vibratory motion can be created using a pneumatic device that slides back and forth; the orientation of the device dictates the vibratory motion. The most popular method is a motor-driven flywheel that is unbalanced. This method requires the use of directional spring members to control the vibratory motion (see Figure 9.1).

Vibratory conveyors can be designed to convey products up inclines and around corners. One of the unique aspects of vibratory conveyors is that they can be designed to fit into spaces that conventional belt conveyors cannot. For instance, vibratory conveyors can be mounted under automated machine tools to catch chips and coolant and convey them out to a conventional chip conveyor. Using a perforated pan, the vibratory conveyor can be used to separate the coolant from the chips. This same idea can

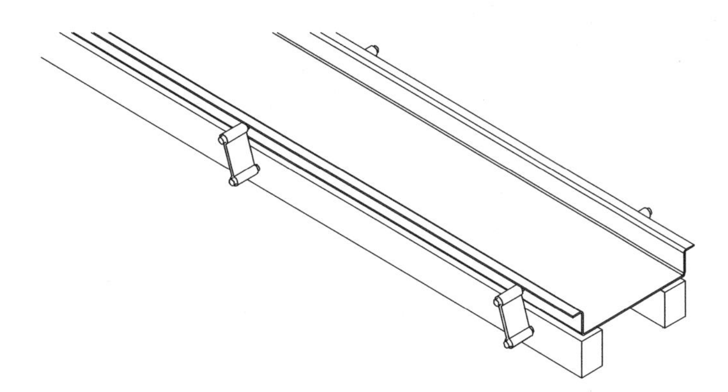

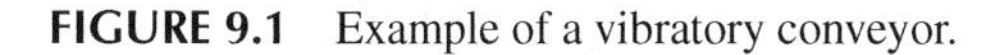

FIGURE 9.1 Example of a vibratory conveyor.

be used for separating any material by size; the smaller product falls through the perforations, possibly into another conveyor, as the larger products are conveyed away.

There are many styles of vibratory conveyors. Some use flat steel or stainless steel tables or pans for the conveying surface. This is by far the most prevalent conveying surface. Using stainless steel for the pan vibratory conveyor is an ideal means of conveying food products. Because there is no belt to keep clean and virtually nowhere for bacteria to hide on a continuous smooth surface, vibratory conveyors are very popular for handling produce and other foods.

Conveyor belting supported in a troughed shape is more popular for non-food-grade applications. The troughed belt variation is used most frequently in applications with aggregate such as gravel, coal, and so on.

One more type of vibratory conveyor offers directional product flow without directional vibration. It uses a biased brush surface that moves the product in the direction in which the brush bristles are pointed. This is useful in handling small, rigid parts and can provide accumulation and singulation of the parts in a single conveyor. Within the table surface, simply arrange the direction of the bristles to create a conveying path.

9.3 AIR CONVEYORS

There are three primary types of air or pneumatic conveyors: bulk, under-supporting, and hanging. Each type has its own applications. Regardless of the type, they all use air to move product, whether through pressure or vacuums. We do not go into great detail about air conveyors because the vast majority of them are custom designed for specific applications.

9.3.1 Bulk

Bulk air conveyors are a reliable, easily maintained, and cost-effective method for transporting many different materials, including powders, plastic pellets, grain, dirt, lightweight parts, and wood chips. There are two primary types of bulk air conveyors: dense phase and dilute phase conveyors. They differ by pressure and the speed of product movement.

Dense phase air conveyors are run by compression and are used for conveying mainly heavier materials. As a result, these convey at a slower rate. Dilute phase conveyors convey by creating a vacuum. This method is much quicker and is used to convey smaller, lighter materials.

Regardless of the type used, when conveying plastic materials in a plastic pipe, static electricity can pose a serious problem. The conveyor manufacturer should have solutions for how to properly ground the equipment so that the static is dissipated and does not become a hazard to surrounding personnel or equipment.

9.3.2 Under-Supporting

Under-supporting air conveyors use a plenum on the bottom with louvers cut in the top to provide directional air flow to lift and move the product. There is virtually no friction between the product and the bottom surface and only minimal friction

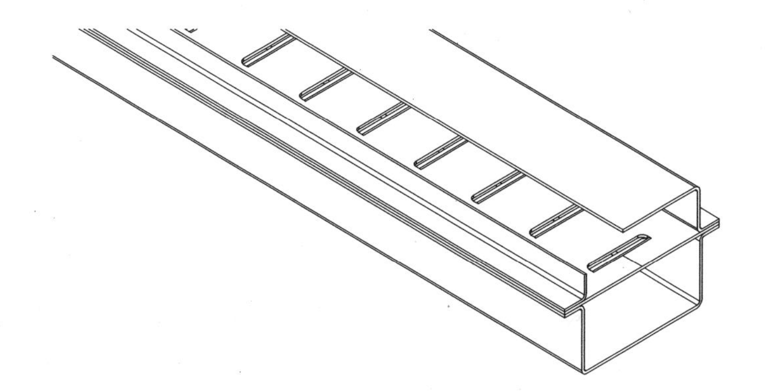

FIGURE 9.2 Under-supporting blower conveyor.

against the side guides. With no moving parts except for the air itself, it allows lightweight product to move very quickly. Back when base cups were still being glued to the bottom of plastic soda bottles, the base cups could be conveyed at speeds too fast to actually see them as they went by. This type of air conveyor is well suited to lightweight products and provides sufficient surface area to lift the product off of the louvers. For obvious reasons, along with normal side guides, top guides are used to keep the product securely on the conveyor (see Figure 9.2).

Moving lightweight products at very high speed is easy to do with an air conveyor, there is one issue to pay attention to when trying to redirect that product. In the above-mentioned base cup application, the base cups only weighed about an ounce, and we used paddle-type gates constructed of ¼" thick welded steel. The vibration for the high-speed impact from the base cups broke all the welds after only a week. The gates had to be redesigned to allow a rounder, sweeping path for the face of the divert paddle to reduce the impact vibration.

The speed of the product can be controlled in three ways. First is through the use of dampers that were mounted on the intake for the blowers; as the air decreased, the product slowed. Second is the use of AC inverters or other electronic speed controls to control the speed of the blower motor. Third, at certain areas where the product always had to be slowed down, such as prior to entering a processing machine, the louvers could be shaped differently to force the air in a more vertical direction rather than the typical horizontal direction.

9.3.3 Hanging

This type of air conveyor is used primarily for transporting empty polyethylene terephthalate (PET) bottles. The bottles are supported by the flanged ring around the neck of the bottle (see Figure 9.3). Typically, polished stainless steel or machined ultra-high molecular weight polyethylene (UHMW)-PE is used to support and guide the bottles. In the case of the hanging air conveyor, the plenum is above the product. The movement is provided by blowers that are mounted on top of the plenum. Because the bottles can move freely side to side, guides are provided to keep them from getting twisted and jammed.

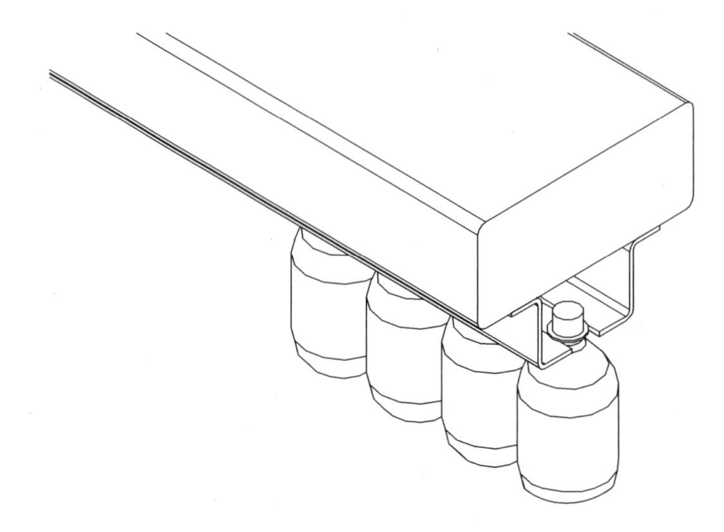

FIGURE 9.3 Hanging blower conveyor.

Product speed is controlled the same way as described for under-supported air conveyors.

9.4 SCREW CONVEYORS

Screw conveyors have been around for thousands of years. In the third century BC, Archimedes designed the first screw conveyor. Screw conveyors use a continuous strip of metal rolled into a helix wrapped around a central shaft or pipe to form a long screw. The concept is that as the screw turns, the product is pushed along by the face of the helix. With each revolution of the screw, the product moves forward the distance of one pitch of the screw.

Frequently, short screw conveyors are mounted on the bottom of large hoppers to meter the flow of product out of the hopper. These are referred to as screw feeders.

The basic screw is very simple: a helix wrapped around a central shaft or pipe. The standard pitch of the screw is equal to the outside diameter. There are a variety of others as well, as detailed in Table 9.2 and shown in Figure 9.4.

There are also screws that have a cone-shaped center and that might have varying screw pitches that are used in screw feeder applications to ensure even product flow through the screw.

Depending on the product to be handled, special features can be added to the screw. The screw can be highly polished to aid in handling sticky product. It can have a plasma coating to increase its hardness to aid in handling very abrasive products.

Screw conveyors obviously consist of more than just a screw. The screw operates either inside a pipe or in a semicircular trough. The trough, which is used for heavy-duty applications, can be U-shaped with parallel vertical sides or it can have flared sides. The U shape is the most popular trough design. The flared trough is popular in applications where the product to be handled is lumpy or stringy. The screw is suspended in the trough by bearings on each end and hanger bearings along the length

TABLE 9.2
Conveyor Screw Pitches and Their Uses

Screw Pitch	Use
Standard pitch	Used for conveying product horizontally or up slight inclines
Half pitch	Used in screw feeders, also for conveying product up an incline
Double-flight full pitch (two helixes on a common shaft)	Used where an even discharge of product is required, typically used only as the last few pitched prior to the discharge
Ribbon flight	Used to convey sticky products or for mixing product
Cut flight	Used where mixing product is desired
Cut-and-fold flight	Used where high levels of mixing product and increased retention time are desired
Varied pitch flight	Used in screw feeders, short pitch below inlet, full pitch beyond to allow product loading to drop

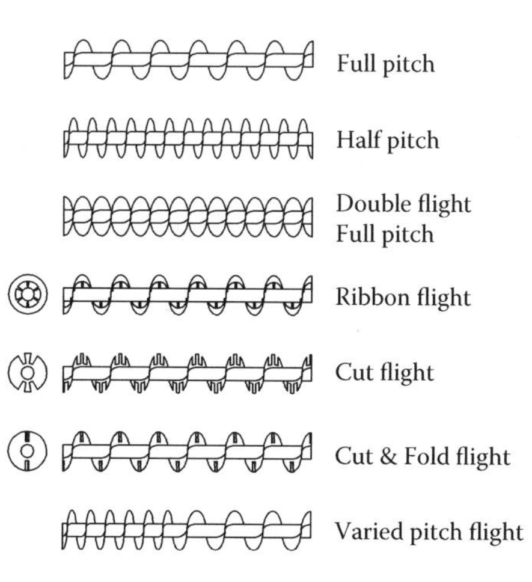

FIGURE 9.4 Various types of conveyor screws.

of the conveyor. The bearings are typically *not* normal ball bearings due to the high probability of contamination. For this application, the bearings can be hard iron, UHMW, Arguto, or ceramic-coated phenolic resin.

For light-duty applications such as small-diameter screws for moving grain, the screw is supported by bearings on either end and then the screw rests against the side of the pipe. The screw can have a UHMW flight edging applied to limit the wear.

The drive for a screw conveyor is a simple arrangement that drives the center shaft or pipe. The drive can be shaft-mounted, or it can incorporate V-belts or chain. Regardless of drive type, the drive is typically located at the discharge end of the screw conveyor. The drive can be mounted at the charge end if access for maintenance is easier at that location.

Depending on the product to be handled, it may be necessary to control the flow of the product into the screw conveyor to prevent it from becoming overloaded or jammed.

FIGURE 9.5 Sand, aggregate, and cement reclaimer.

Screw conveyors can also be used for material separation. For instance, one manufacturer, Vince Hagan Co., offers a concrete separator. Old concrete is dumped as it is washed out of the mixer truck into a screw conveyor. More water is added through sprayers, and a pair of screws carry it over various-sized screens to allow the aggregate to fall out, then the sand, and then, finally, the water and cement mixture. The water and cement mixture is released into containment ponds, where the cement and other constituents settle out and the water is reused to make more concrete (Figure 9.5).

9.5 BUCKET ELEVATORS

Bucket elevators are designed to convey product in a primarily vertical direction. Bucket elevators come in a variety of sizes to handle products that range from dry, dusty powders to heavy iron ore or aggregate. Bucket elevators work best with dry materials, although some manufacturers have developed designs that will work acceptably with wet materials. The key to conveying wet materials is that they cannot be sticky or tend to buildup or collect in the buckets.

Product is fed to the bucket elevator through a hopper. Buckets or cups dig into the product, convey it up, and dump it out at the discharge. Figure 9.6 shows several typical configurations for bucket elevators.

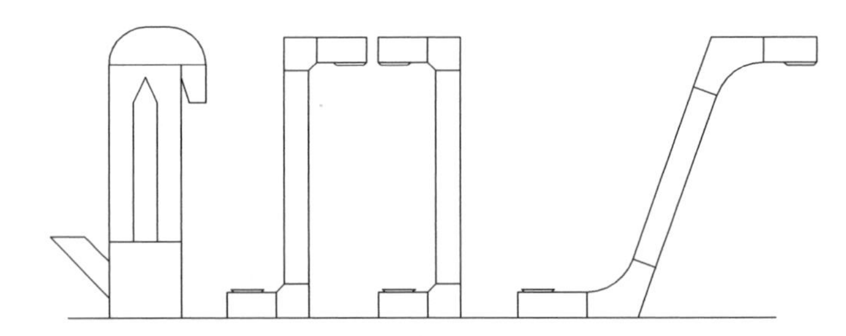

FIGURE 9.6 Examples of bucket elevator arrangements.

Bucket elevators are built in a variety of configurations. The basic designations have to do with how the product is discharged and how the buckets are mounted. The two types of discharge are centrifugal and continuous. The mounting of the bucket comes in a variety of configurations. The bucket can be mounted to a flat belt, it can be an integral part of a plastic belt, it can be mounted rigidly on one or two chains, or it can be suspended between two chains. First, we look at the two discharge types.

9.5.1 Centrifugal Bucket Elevators

Centrifugal bucket elevators operate at higher speeds to throw the product out of the buckets using centrifugal force as the buckets pass over the drive sprocket or pulley. To prevent interference between discharging product and adjacent buckets, the buckets are spaced a little further apart than in a continuous elevator. Centrifugal bucket elevators are frequently used to move free-flowing products such as grain, sand, minerals, sugar, and aggregates.

9.5.2 Continuous Bucket Elevators

Continuous bucket elevators can handle the same products as a centrifugal bucket unit but are designed to handle products that are easily damaged, sluggish, or abrasive. They also work well with product that may have large lumps in the product mix. They can operate at lower speeds to minimize breakage of friable product. Materials that should not be aerated should be avoided. The buckets are typically spaced closer together. As product is poured out, the product flows over the back of the previous bucket. Frequently the backs of the buckets have flanges on them to form a chute to guide the product to the discharge.

9.5.3 Rigid Buckets

Rigid buckets, as stated earlier, are mounted to a belt or chain or can be an integral part of a plastic belt. The buckets can be constructed from steel, stainless steel, aluminum, or molded plastic, typically nylon. The buckets are bolted through the flat belt using flathead screws and are evenly spaced over the length of the belt. Likewise, the buckets can be bolted to special attachment links on a roller chain. With chain-mounted buckets, it is critical that the chains be kept consistently in sync with each other. Otherwise, the buckets can get twisted and this prevents them from navigating the end sprockets properly. Chain elevators typically are used for handling nonabrasive materials, whereas belt elevators will handle more abrasive material as long as there are no slivers or sharp objects that will damage the belt. Multiple staggered rows of buckets can be used for wider bucket elevators.

9.5.4 Suspended Buckets

Suspended buckets are typically mounted so that they hang freely between two chains. Some manufacturers do offer their bucket suspended between two specialty belts. Either way, the suspended bucket will retain its contents until it is tipped over.

This allows the elevator to have multiple discharge points. This is accomplished using a mechanism that can be triggered to tip the buckets as they go by. This cannot be used as a sorter for multiple products but can be used to balance the product load among multiple discharge points. The buckets are typically made from the same materials as the rigid buckets.

9.6 CHIP CONVEYORS

Chip conveyors are used to remove metal chips, stamping scrap, and turnings from a press or machine tool and deposit them either in a scrap hopper or onto a common conveyor that collects chips from multiple machines. Because of their all-metal construction, materials that are too hot for typical conveyors can be handled by most chip conveyors.

There are some ancillary machines that accompany most chip conveyors (such as filters and wringers) that are not discussed here.

9.6.1 Hinged Steel Belt Conveyors

Hinged steel belt conveyors, sometimes referred to as apron conveyors, are useful for a wide range of horizontal and inclined chip conveyors. Hinged steel belt conveyors handle metal chips or stringy metal turning chips.

9.6.2 Drag Flight Conveyors

Drag flight conveyors are used for conveying broken or small metal chips. They are not recommended for stingy turning chips or other materials that bunch because they will cause jamming for the drag chain. Drag flight conveyors are available in the same configurations as the hinged steel belt conveyors. Typically, the conveyor is a rectangular trough with one or two strands of chain that drag the flights along the bottom of the trough. The flights can be equipped with scraper blades to make better contact with the bottom of the trough.

9.6.3 Magnetic Chip Conveyors

Magnetic chip conveyors are used for conveying the chips and shavings from any ferrous materials. The conveying surface is a sealed surface with magnets mounted below the surface on a recirculating chain with no exposed moving parts. The chips are attracted to a magnet and follow it until the magnet is moved away from the surface. One of the advantages is that coolant is easily drained from the conveying surface.

9.7 AUTOMATED ELECTRIFIED MONORAIL

An automated electrified monorail (AEM) consists of individually powered carriers that get their power from a common track-mounted bus bar arrangement. In many applications, AEMs have replaced slower overhead P&F systems.

With each carrier being independent, there is no chain that has to be driven, routed between points, or lubricated and maintained. The carriers can move bidirectionally and can also travel at various speeds. In production areas where the system interfaces with other equipment, the carriers can move at, say, 20 FPM. In long-distance transportation areas, the speeds can be increased to, say, 300 FPM. The bidirectional feature is useful when positioning is critical. If the carrier must back up for accurate locating, it has that capability.

As with the power and free that was discussed in Chapter 8, "Overhead Chain," there is a variety of accessories such as switches and merges that are typically managed by the electrical control system. These allow carriers to follow multiple paths depending on the product they are carrying. The controls system can either use registers in the programmable logic controller (PLC) or onboard identification such as radio frequency identification (RFID) tags and tag readers at the switches.

From a maintenance standpoint, the AEM system allows individual carriers to be removed and serviced. With multiple drives, there is no longer the concern that the primary chain drive will fail and bring the entire system to a halt.

9.8 ASSEMBLY PLATFORM CONVEYORS

Assembly platform conveyors (APCs) are ideal for large production assemblies such as motor vehicles. APCs are frequently referred to as skillet conveyors. With the integration of vertical lifts, turntables, and other custom-designed components, the product being conveyed is oriented in the optimum ergonomic position. This enhances worker comfort, safety, and productivity.

APCs consist of a large platform that will have the lift or turntable mounted on it. The platform is actually moved by a large conveyor that is mounted below the floor. APCs typically operate at 10 to 25 FPM. The platforms are kept close together, effectively creating a moving floor.

CASE STUDY – PICTURE TUBES

Once I was handed a project for handling the back half of CRT picture tubes for televisions. The customer was going to be making the largest flat-screen picture tubes on the market at the time, 38″. I know that the whole picture tube idea really dates this, but the thought process behind the conveyor design shows how new designs are born. This case study is in Chapter 9, "Miscellaneous," because you will see that this is definitely not a typical, standard conveyor.

The material to be handled were rectangularly shaped glass funnels ranging from 25″ to 38″ picture tubes. Therefore, everything had to be adjustable. We were to convey the funnels from a power and free conveyor through five robotic grinding and polishing stations and three rinse hoods and deliver them to a drying oven.

The customer had previous experience with other production lines for smaller funnels, so we took advantage of their knowledge. It is sometimes hard for some engineers to admit they don't know everything, but good ones will accept good ideas from anywhere. I frequently say, "Half of being smart is knowing what you're dumb at."

The funnels were going to go through five robotic grinding and polishing operations as well as three rinse hoods to wash off the finishing compounds. Therefore, the conveyor had to have a trough under it to catch the water and direct it to drains to be filtered and recycled. Some of the processes used deionized water, so everything had to be stainless steel. The deionized water would quickly corrode any mild steel. Also, all aluminum parts had to be anodized to protect them. We used different colors for the anodizing so that similar parts were obviously different.

The first challenge was to determine the primary conveying surface. The customer had previously used two large urethane belts set approximately 12″ (305 mm) apart for smaller lighter weight products. It was quickly determined that this would not work due to the 40-pound (18 kilos) weight of the largest funnel. We were looking for something that would have a soft conveying surface so as not to chip the glass funnels, yet strong enough to convey multiple funnels distances up to 40′ (12.2 m). Additionally, whatever we used had to be able to be guided to keep the belt straight over the length of the conveyor. We selected a 50-mm-wide BRECOflex brand timing belt that was stainless steel cable reinforced with a center V-belt for tracking purposes. The belt would be vulcanized to eliminate the metal lacing that could chip the funnels. We would use two belts as well to provide stable conveying surface.

At the drive end of the conveyor, a robot would be picking up the funnel to place it in a grinder or polisher. We had to develop a mechanism to square the funnels to the conveyor to ensure they had the proper orientation before they were placed in one of the finishing machines. In Figure 9.7, you can see two L-shaped devices on the end of the conveyor. As the funnel would approach, they would accept the funnel and then the driving force of the conveyor belts would push the funnel squarely into the devices.

The next challenge was how to accumulate the product on a belt conveyor. With two strands of belts, we had to develop a lift that would fit in between the two belts. The lifts had to lift smoothly and as level as possible so that the funnels would not skew when lifted up or set down. This level requirement eliminated most typical pneumatic lifting devices. We were left with either a parallelogram lift or a scissor lift mechanism. After some testing, it was determined with input from the customer that the straighter lift of the scissors lift mechanism was superior. In Figure 9.7, you can see the lift between the two belts. We also added two urethane pads along the top of the lift to protect the funnels.

The one challenge both mechanisms had was that the pneumatic actuators were erratic and bouncy. To eliminate these negative aspects, we determined that each lift would have to have an air-over-oil tank. This would allow each lift to actuate and move as though it was hydraulic, yet still be pneumatically actuated. In Figures 9.8 and 9.9, you will see the air-over-oil tanks all along the conveyor, one next to each lift. For the convenience of the maintenance people, we mounted the solenoid valve directly on the back of the tank bracket, so it became a single assembly.

FIGURE 9.7 Conveyor drive end.

FIGURE 9.8 Robot picking up a funnel.

FIGURE 9.9 View down conveyor line.

Another challenge was how to consistently sense the funnels to actuate the lifts. We needed something fool proof. On typical package conveyors, if a box or totes runs into another because a sensor didn't operate properly once, it's no big deal. With the funnels, there was absolutely no allowance for contact between two funnels. We needed something that could sense or see the funnels whether they were wet or dry. Due to the reflectivity of water on a glass surface, photo-eyes were ruled out. We finally developed a sensor roller arm that had a urethane roller on a counterweighted arm. As there would be a lot of garnet polishing dust and other abrasives in the air, we opted to use two stacked agricultural bearings with labyrinth seals to support the sensor rollers. The sensor rollers had a proximity sensor that would be triggered when a funnel hit the roller. The roller's arm had to be designed so that the proximity sensor remains triggered even when the funnel was lifted up of the accumulation lift.

Figure 9.8 shows a Fanuc robot about to set a completed funnel down on far conveyor while picking up a new funnel from the closer conveyor.

You will notice in Figures 9.7 and 9.9 there are large white rollers on the end shafts. These are there to support the outer edges of the funnels as they transition across the end of the conveyors.

This design project was not completed quickly and not without mistakes and missteps. From start to finish, it took two years. We spent a great deal of time working with the customer's engineers, maintenance people, and production line personnel. We tested every new concept to ensure a robust design. When we completed, the customer had a very solidly built, robust conveyor system that would operate for years with minimal maintenance.

9.9 QUESTIONS

1. What would be the drawback to using a straight chute in a distribution warehouse that handles computers versus one that handles clothing?
2. List potential products that could be handled by a screw conveyor.
3. Which conveyor type would be a better choice for moving dry cement up 40' vertically: bucket elevator or screw conveyor and why?
4. If products are coming down a chute too quickly, which of these is not a method of slowing the product?
   a. Mount belting to the bottom of the chute to create a rough spot.
   b. Hang a vinyl strip curtain across the chute to create extra drag of the product.
   c. Mount belting to the side of the chute to create a rough spot.
   d. Raise the discharge end to reduce the angle of the chute.
5. True or False: a vibratory conveyor with a perforated conveying surface can be used to separate metal chips from the cutting fluid under a machining center and still move the chips out from under the machine.
6. Very dry, very fine sand tends to flow like water. What features of a screw conveyor would impact its effectiveness in conveying this material?
7. What is a safety concern when conveying plastic pellets in a plastic pipe using air?
   a. Controlling the speed of the pellets in the pipe
   b. Rigid support structure for the pipe
   c. Grounding the pipe to eliminate static electricity buildup
   d. All of the above
8. If you had an application that required a P&F-type conveyor, but the carriers had to move independently and at higher speeds than typical P&F, what conveyor type would you select?
9. If you wanted a bucket elevator that had multiple discharge points, what kind of bucket elevator would you select?
10. Select the correct chip conveyor to handle aluminum chip from a machining center that can be long and stringy.
    a. Hinged steel belt conveyor
    b. Drag flight conveyor
    c. Magnetic conveyor
    d. a or c

10 Rate Calculations

Rate is a term bantered about frequently when dealing with any material handling system. Conveyors are frequently applied when other types of material handling equipment cannot achieve or maintain a consistent rate.

Merriam-Webster defines rate as "a quantity, amount, or degree of something measured per unit of something else." In the case of material handling, rate is a quantity of product: cartons, bags, or tons measured per unit of time, such as cartons per minute, bags per minute, or tons per hour.

Let us continue by defining some other important terms. Regarding carton dimensions, *height* is the vertical dimension of the product measured in inches, feet, or meters. *Length* is the dimension of the longer side measured in inches, feet, or meters. *Width* is the dimension of the shorter side, also measured in inches, feet, or meters. *Weight* is the overall weight of the product measured in pounds or kilograms.

If you are not familiar with the following terms, please refer to the Glossary.

- Live load
- Design capacity
- Average capacity
- Maximum capacity
- Speed
- Rate capacity
- Gap: Head to head

10.1 UNIT HANDLING

When dealing with conveyor rates, we typically are concerned with "average" product sizes because most conveyor systems handle a wide variety of products. Just think of all of the different types of luggage that an airport conveyor must handle or the wide array of box sizes that flow through a department store distribution center. The only way to truly analyze a system is to work with the average product size, keeping the minimum and maximum sizes in mind as well. As we go through the calculations, we have segregated the formulas so that the metric version is on the left and the imperial version is on the right. In many cases, the formula is generic, so only one is shown; however, you do need to pay attention to units of measure.

10.1.1 Flat Belt Conveyors

The minimum speed a conveyor can run at is determined by the speed of the conveyor or process feeding it as well as the speed of the conveyor or process that the conveyor is feeding. The minimum speed is the case speed (CSPD) measured in

DOI: 10.1201/9781003376613-10

meters per second (m/sec) or feet per minute (FPM). It is based on the average product length (PL) in meters or feet and the product rate measured in cases per minute (CPM) as shown:

$$CSPD = PL \times CPM$$

Belt conveyors, which have positive control over the product to be conveyed, can run at speeds closer to the minimum speed. For instance, a rate of 20 CPM and an average carton length of 16 in. gives a CSPD of 26.7 FPM. In imperial units of measure, this is typically rounded up to the next increment of 10, or in this case 30 FPM.

Roller conveyors, due to inefficiencies in gripping and controlling the product, typically run faster to ensure proper throughput. Typically, the conveyor is designed to run 20 percent faster than the minimum speed. In the preceding example with a result of 30 FPM, adding 20 percent brings it up to 32 FPM, so the conveyor would be designed for 40 FPM.

CSPD is not the only criterion used in determining conveyor speed. Typically, the speed selected for a conveyor is higher than the minimum required to ensure that the design capacity is met. A variety of other issues plays into the speed selection, such as the speed with which the conveyor is being fed product or the speed of the equipment into which the conveyor is feeding. Higher speeds are also used to compensate for the irregularities of manual loading, points where two conveyor lines merge, and to minimize any bottlenecks in the system.

For instance, in a distribution center, in the pick modules, an operator puts boxes on a belt conveyor. At the beginning of the system, the number of boxes on the conveyor is very low, so the obvious choice would be to run the conveyor slower to minimize wear and tear and also save electricity. The problem with that way of thinking is that as the operator works his or her way along the conveyor, putting boxes on, the boxes already on the conveyor could be in the way. The operator then has to stop and wait for the boxes to move out of the way. That is why most picking conveyors run at 0.6 m/sec (120 FPM) or faster, rather than just 0.2 m/sec (30 FPM) as would be determined strictly by the CSPD.

Let us go back and look at the CSPD calculation and how it can be applied in a system. In many conveyor systems, the product needs to be spaced out or gapped. One of the simplest ways to accomplish this is with two belt conveyors. The first would be running at the CSPD. Doing this bunches the product together with little to no gap between them. The second belt would be running faster and thus pulling a gap between the products. The speed of this second conveyor would be calculated in a similar manner to the CSPD. The only difference is the addition of the gap to the PL.

$$SPD = (PL + Gap) \times CPM$$

An example would be if we had a PL of 610 mm (24 in.) and a rate of 25 per minute, the minimum speed would be 0.3 cm/sec (50 case feet per minute [CFPM]). To add a 230-mm (9-in.) gap between the products as shown in Figure 10.1, our speed would be 0.4 m/sec (70 FPM).

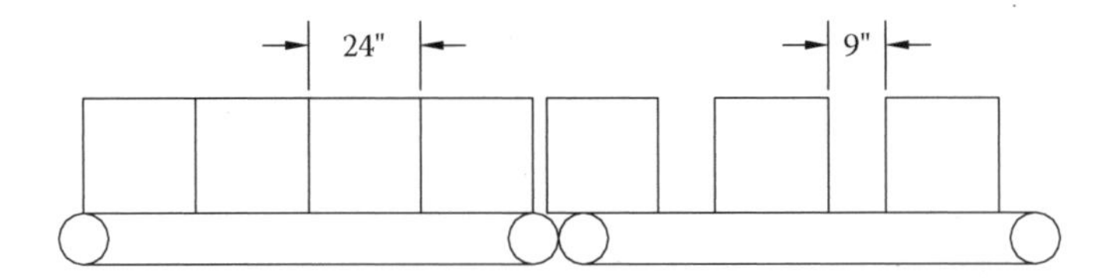

FIGURE 10.1 Using two belt conveyors to generate a gap.

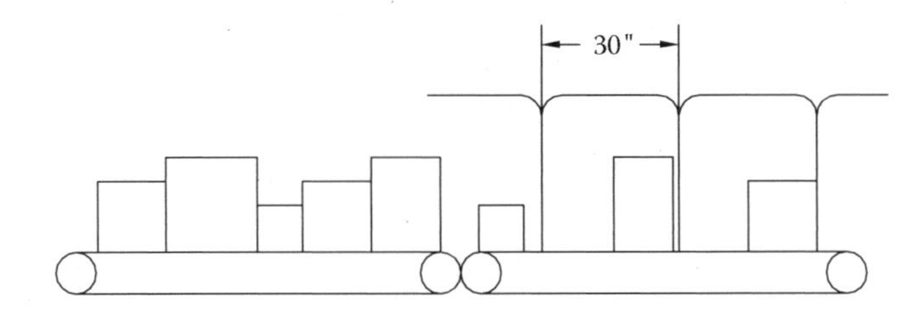

FIGURE 10.2 Using two belt conveyors to generate a product window.

Similarly, there is occasionally a requirement to produce "windows" around the product. This concept occurs in full-case pick modules or when merging product together to reduce or eliminate jams. Window length (WL) is not dependent on product length, but the window must be large enough to accommodate the longest product. This speed calculation is just as simple.

$$SPD = WL \times CPM$$

A window is like a space reservation on a conveyor onto which a product will be placed or is already on, as shown in Figure 10.2. Typically, the product is placed with the leading edge at the front of the window, but in reality it can be anywhere in that window. Windows will become important when we discuss sorters later.

As was mentioned in Chapter 4, "Belt Conveyors," the slave-driven power feeders on inclined belt conveyors typically run 5 percent slower than the incline that drives them. Conversely, if the belt is a decline, then the slave-driven unit runs 5 percent faster. The reason for this speed differential is to prevent the product from interfering with each other. There is a very simple formula to ensure that the speed difference is sufficient using the average product height (PH) and the angle of incline (α; see Figure 10.3).

$$\tan(\alpha) \times PH < .05 \times PL$$

If the speed difference is something other than 5 percent, then simply substitute the new percentage value for the .05.

Belt turns present a few issues that must be considered as well. Belt turns are typically designed based on a centerline speed (SPD_{CR}), which is the speed at the centerline of the effective or usable width. Because the product follows the arc of the curve,

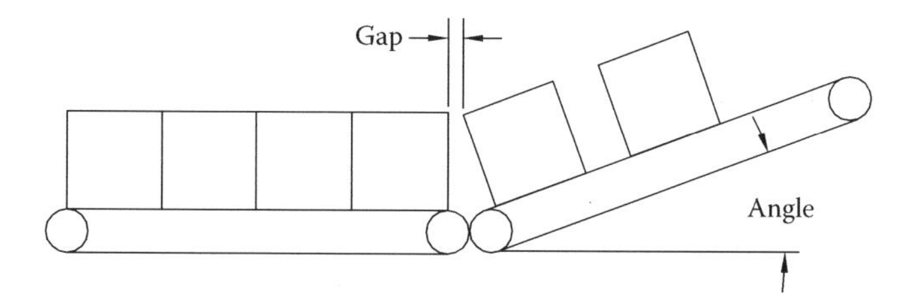

FIGURE 10.3 Speed differential required at power feeders to prevent product interference.

the gap between the products closes toward the inside of the curve. Some integrators, in an effort to minimize risk and engineering time, will specify the curves based on the speed at the inside radius (SPD_{IR}) equal to the speed of the conveyor feeding the curve.

$$SPD_{CR} = \frac{SPD_{IR} \times CR}{IR}$$

To determine if the product is properly spaced, the formula below uses the PL, average product width (PW), and the centerline radius (CR) of the turn.

$$Gap = \frac{SPD \times PW}{2 \times CR \times CPM} - PL$$

This assumes that the product is centered on the curve. Frequently, this is not practical so the alternate way to look at this is if all product is justified to the inside radius (IR) of the curve.

$$SPD_{IR} = \frac{SPD_{CR} \times IR}{CR}$$

$$Gap = \left(\frac{SPD_{IR}}{CPM}\right) - PL$$

The calculations for the belt turns are predicated on the assumption that you have already done the calculations to ensure that the curve is properly sized for the range of products to be handled.

10.1.2 Live Roller Transportation

Up to this point, we have discussed almost exclusively flat belt conveyors. We move next to the live roller conveyors. In discussing the minimum conveyor speed, earlier we touched on an important point that bears repeating: A roller conveyor does not maintain a solid grip on the product being conveyed. Therefore, there is some inefficiency that needs to be compensated for. Thus, we add 20 percent to the minimum speed. This 20 percent can be reduced to as little as 5 percent if the conveyor has some special coating on the rollers that improves their grip. For the sake of this text, we stick with the 20 percent. As you will notice, we take the formula for the minimum speed for a belt conveyor and add 20 percent by multiplying by 1.2.

$$SPD = PL \times CPM \times (1.2)$$

Just as the speed of the straight roller conveyor is higher than the straight belt conveyor, the same holds true for the live roller curve. Use the same formulas to determine the speeds of the curves and add 20 percent (or whatever value you have selected). Keep in mind that depending on the conveyor type with which you are working, the drive for the curve may be the same as the straight; the question then becomes whether the curve is driven on the inside or outside of the curve rollers. Each manufacturer's roller geometry is different so it is important to know how fast the curve will be running as compared to the straight conveyor powering it.

10.1.3 Live Roller Accumulation

We have looked at the rates for strictly transportation conveyors thus far. Now we discuss the rates for accumulation conveyors. Because of the length of many accumulation conveyors, the rate at the infeed or charge end is completely different than the rate at the discharge end. As such, we look at them separately, keeping in mind that for any given conveyor there can only be one speed, so we go with the faster requirement.

In Chapter 6, "Powered Conveyors," we discussed the various types of sensors – mechanical, pneumatic, and electronic – as well as the various types of drives. They all have in common a way to sense the presence of a product and the ability to add and remove drive to power the rollers and thus move the product.

Because of this accumulating action of the rollers removing the drive to the rollers in the previous zone, the products end up with gaps between them equal to at least the zone length. This results in a throughput of approximately half of what it would be if the product was back to back. Because the product is moving along in a separated manner, this is referred to as singulated.

Due to the reduced throughput of a conveyor because of this spacing, there are two additional types of accumulation: slug and dynamic.

Slug accumulation is basically turning a section of accumulation conveyor into a transportation conveyor. This is done through the use of pneumatic or electronic controls, which basically overrides the sensing capabilities. When appropriate, the control system can turn the accumulation function back on. Most mechanical sensor systems do not offer slug operation.

Dynamic or hybrid accumulation is achieved when a solid slug system is used at a reduced drive strength, causing the conveyor to function as a low-pressure live roller transportation unit. Dynamic accumulation is typically used for leading into an induction system such as prior to a sorter (see Table 10.1).

As we discussed before, a roller conveyor is not 100 percent efficient when moving product; therefore, we add 5–20 percent for a given rate. Accumulation conveyors

TABLE 10.1
Accumulation Types and Uses

Application	Charge End	Discharge End
Lines feeding a manual operation	Slug	Singulated
Recirculation lines on a sorter	Slug	Singulated
Lines feeding a merge	Singulated	Slug
Lines that feed another accumulation conveyor	Singulated	Singulated
Induction to a sortation system	Dynamic if under 50-ft. long Singulated over 50-ft. long	Dynamic

are less efficient than transportation conveyors due to the inherent gapping of the product. The efficiencies of the various types of accumulation conveyors are as follows:

- Singulated accumulation: 50 percent
- Slug (transportation) mode: 95 percent
- Dynamic accumulation: 76 percent

To determine the speed of the accumulation conveyor, we start with the familiar minimum speed.

$$CSPD = PL \times CPM$$

We then compensate for the efficiency of the accumulation conveyor type.

$$SPD = \frac{CSPD}{Efficiency}$$

Depending on the roller surface used, the efficiency at lower speeds could be better than the efficiency at higher speeds. Singulated accumulation running at 100 FPM has 50 percent efficiency, but at 300 FPM the efficiency may drop to 46 percent. This is because when a roller is suddenly driven when a zone is reenergized, the rollers will tend to spin a little under the product prior to its beginning to move. The surface preparations that could overcome this lower efficiency are typically more costly than simply running the conveyor a little faster.

10.1.4 Merges

The rate through merge points is the next area we examine. A variety of merge types needs to be considered: angled merges, perpendicular merges, and wide-bed merges. Whenever you are trying to merge two or more conveyor lines, make sure that the take-away conveyor, the conveyor that ultimately will handle all of the product, is running fast enough to handle it. This speed is more than the minimum conveyor speed based strictly on product size and product per minute. When merging lines together, the product really needs to be moving in windows, and we already discussed how to determine speed based on windows. The challenge is sizing the window.

To properly size the window, we have to look at the merge and the angle at which product will be entering. The typical formula for a merge window is based on the maximum product width (MPW) in feet (rather than the average product width) and the angle of the merge (A).

$$Window = \left(\tan\left(A\right) \times MPW\right) + PL$$

This, of course, only works for angled merges, because when there is a perpendicular or 90-degree merge, the tangent of the angle becomes infinite. For 90-degree merges, we use a slightly different formula. One of the challenges of a 90-degree merge is that as the product transitions from one conveyor to the other, it must change orientation. Typically, a turning post is provided to minimize the drag of the product as it navigates the corner. This reorientation decreases the efficiency of the conveyors and plays into the following formula shown below. Here we use the maximum product length (MPL), and the speed of both the take-away (SPD_t) and the infeed (SPD_i) conveyors.

$$Window = \frac{MPL \times SPD_t}{SPD_i \times Efficiency}$$

Depending on the roller surface used, the efficiency through a turning post type 90-degree transfer is 50 percent. If a turning post is not used, this number will shrink; conversely, if the rollers have a knurled surface or urethane coating, the efficiency will increase.

Another way to increase the efficiency through a 90-degree transfer is to break up the take-away line so that the conveyor upstream of the merge point can be stopped and trains or slugs of product can be released at once. This allows a slow or sluggish product to be pushed through the transfer point by the units that follow.

On live roller systems, there is one additional control method at a merge point: a traffic cop. A traffic cop is a mechanical device that holds back product on one line while product on the other is going through. As soon as the active arm returns to its rest position, product on the other line will begin to move. This works very well in areas of limited product flow because it requires no electronic or pneumatic controls. The maximum recommended conveyor speed when applying a traffic cop is 120 FPM. The maximum rate through a traffic cop is about 12 CPM based on an average length of 24 in. Otherwise, the product on one line is too close together, and the traffic cop will never return to its rest position, so the other line will never have the opportunity to be released.

When merging two parallel lines, the most popular merge used is the wide-bed merge. This is basically a belt or live roller conveyor that is as wide as the two incoming conveyors, including the space between them. To move the product from one side to the other, either a fixed deflector or a vertically mounted belt conveyor is used.

When applying a wide-belt merge, one or both of the lines feeding the merge must have a stop of some kind prior to the merge. The stop can be either a pop-up stop or a brake belt conveyor. If the exit conveyor is in line with one of the infeed conveyors, the other infeed always requires a stop or brake belt. If the exit is centered, both lines need a stop or brake belt. In the latter arrangement, the straight-through line runs continually and the other keeps the product stopped until the control system determines that there is a sufficient gap into which the product can merge. This works well when the product being merged arrives at a fairly low rate.

Most systems that use a wide-bed merge use stops or brake belts on both incoming conveyor lines. In this situation, product is accumulated behind one of the stops while the other line is flowing. After a while, or when the waiting line becomes full, the running line is stopped and the other is allowed to flow.

The choice between the fixed deflector and the vertical mounted belt is based on the rate requirements through the merge area. Whenever product is forced to move side to side on a conveyor, there is a loss in efficiency. The fixed deflector tends to reduce the products' speed, thus increasing the inefficiency, but the cost is considerably less than the vertical belt. The vertical belt is designed to match the conveyor speed to aid in minimizing any inefficiency. The speed of the vertical belt SPD_v is determined based on the straight conveyor speed (SPD) and the SIN of angle of the deflection (α).

$$SPD_V = \frac{SPD}{SIN_\alpha}$$

Another merge type involves a wide-bed merge with a movable center deflector. In this case, both conveyors can flow straight through the merge area. The deflector offers a merge capability when required. Both incoming lines are equipped with a stop or brake belt to prevent product from entering the merge while the arm is changing positions. Although it seems as though this would work for a simple sortation point, it is not recommended unless the rate is very low due to the time required to shift the deflector from one side to the other. It would require a very large gap between products.

A variety of other merge conveyors is available (such as live roller merges, belt merges, and saw-tooth merges), but they all work in the same basic manner and require the same considerations, as previously discussed. The advantages of these other conveyors are that they can increase the efficiency through the merge and thus increase the possible rate.

An alternative to the wide-bed merge is the use of a pusher that would quickly push the product from one line to the other as soon as there is a sufficient gap. This, of course, allows for a smaller gap between product into which the merged product can fit. However, this method is far less popular than the wide-bed merge.

Accessories in a system can affect the system rate. Transfers and turntables are two such accessories that offer system flexibility but have an adverse effect on rate. This is because the product rate through a transfer or turntable is lower than the normal straight-line rate.

$$Rate = \frac{SPD}{PS}$$

First, let us look at transfers. The product spacing (PS) measured in centimeters (inches) depends on several factors including main line conveyor speed (CS) and transfer speed (TS) both measured in meters per second (m/sec), maximum product length (MPL), and conveyor width (CW), both measured in centimeters (inches).

First, we determine the time constant (T) for the conveyor. This includes 1.0 sec for clearance and 1.6 sec for activation time (reaction 0.3 sec + acceleration 1.0 sec + deactivation 0.3 sec).

$$T = \left(\frac{CW}{TS \times 100}\right) + 1.0 + 1.6 \quad T = \left(\frac{CW \times 12}{TS \times 60}\right) + 1.0 + 1.6$$

PS is then determined by:

$$PS = \left[T \times CS \times 1000\right] + MPL \qquad PS = \left[\left(\frac{T}{12}\right) \times CS\right] + MPL$$

This calculation determines product spacing that would allow the system to operate without stopping the pallets except to position them on the transfer. However, the system should be designed so that there is the opportunity to stop oncoming product prior to the transfer. This ensures that there is no chance of a product running into the raised transfer.

Turntables require a slightly more complex calculation. PS depends on several factors, including main line conveyor speed, turntable speed (TtS) measured in revolutions per minute, turntable diameter (TD), and MPL, both measured in centimeters (inches).

First, we determine the time required for the turntable to turn and return (TtT).

$$TtT = \frac{60}{TtS \times 2}$$

Next, we determine the time constant (T) for the conveyor. This includes 1.0 sec for clearance and activation time (reaction 0.3 sec + turn time [TtT] + deactivation 0.3 sec).

Metric

$$T = \left(\frac{TD}{CS \times 100}\right) + \left(\frac{MPL}{CS \times 100}\right) + 1.0 + \left(0.3 + TtT + 0.3\right)$$

Imperial

$$T = \left(\frac{TD \times 12}{CS \times 60}\right) + \left(\frac{MPL \times 12}{CS \times 60}\right) + 1.0 + \left(0.3 + TtT + 0.3\right)$$

PS is then determined by:

$$PS = [T \times CS \times 1000] + MPL \qquad PS = \left[\left(\frac{T}{12}\right) \times CS\right] + MPL$$

This calculation determines product spacing that would allow the system to operate without stopping the pallets except to position them on the turntable. However, the system should be designed so that there is the opportunity to stop oncoming product prior to the turntable. This ensures that there is no chance of a product running into a turntable that is not in position.

The final type of unit handling conveyor we look at is sorters. There are a variety of sorters, but they all have the same rate and speed calculations.

$$Rate = \frac{SPD}{PL + Gap}$$

Therefore, conversely

$$SPD = Rate \times (PL + Gap)$$

These are the same basic formulas put forth earlier. The challenge becomes determining the required gap based on the type of sorter being used. Table 10.2 provides a list of sorter types and their respective gap calculation.

As we mentioned earlier, there is always the efficiency of a conveyor, or in this case diverter type, to consider. In many of the preceding formulas, there is a multiplier on the end such as 1.33, 1.5, or 1.1. These are the efficiencies of that divert method. The original equipment manufacturer may use a slightly different factor based on its specific design, actual product testing, or both.

Thus far, we have discussed belt conveyors and roller conveyors, which are typically used for package handling. There are a variety of unit handling conveyors that are based on carriers. These use the same speed formula as if you are using windows. In this case, the window length is the distance between the same point on the carriers (CarCtr) measured in feet.

$$SPD = CarCtr \times CPM$$

The final type of unit handling conveyor we discuss is the tabletop chain (TTC) conveyor. TTC conveyors act very similar to flat belt conveyors in that a product will move at the same speed as the chain until an outside force acts on it. The difference is that TTC is specifically designed to slide underneath the product. This slipping is the reason that typically an efficiency factor of 20 percent is added to the chain speed.

$$SPD = PL \times CPM \times (1.2)$$

This formula works well when the product is single file. When we are dealing with accumulation situations, we need to look at the shape of the product. For square or rectangular products that pack together neatly, the speed can be divided by the

TABLE 10.2
Sorter Rate Calculations

Sorter Type	Gap Calculation
Pusher on live roller or belt conveyor	$Gap = \frac{CT \times FPM}{60} \times 1.33$
Paddle or divert arm on live roller conveyor	$Gap = Sin(\alpha) \times MPW \times 1.5$
Pop-up wheel divert in live roller conveyor	$Gap = Sin\ 30^{\circ} \times MPW \times 1.33$
Pop-up belt transfer in live roller conveyor	$CT = \left(\frac{ConvWidth \times 60}{FPM_T}\right) + 1.0 + 0.4$ $Gap = \frac{CT \times FPM}{60}$
Pop-up wheel divert in belt conveyor	$Gap = Sin\ 30^{\circ} \times MPW \times 1.1$
Sliding shoe sorter	The minimum gap is determined by the manufacturer

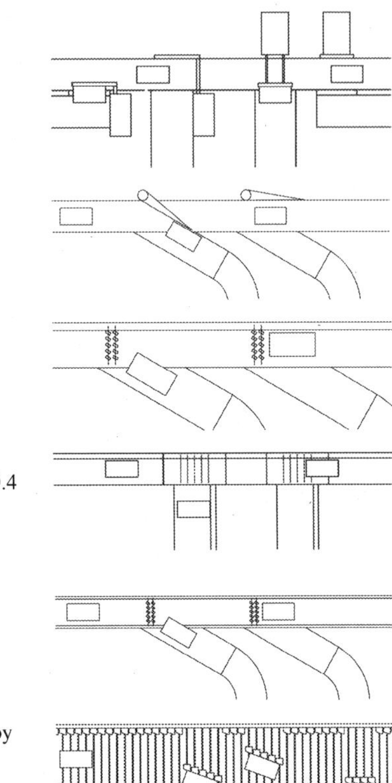

Note: CT = cycle time in seconds.

number of products across the width of the chain. If the product is 3-in. wide and the chain is 12-in. wide, then we can divide the speed by four.

$$\frac{12''}{3''} = 4 \therefore \frac{\text{FPM}}{4}$$

Round product, on the other hand, is not as simple. Round product nests itself together. Even if the product is 3 in. in diameter and the chain is 12-in. wide, the product will not be in nice rows of four abreast. Instead, the round containers will nest together and create alternating rows of three and four. Here, we use a formula to determine the number of products in a given area. PPSM is product per square meter and PPSF is product per square foot. In both cases, D is the product diameter measured in millimeters or inches, respectively.

$$PPSM = \frac{1.15 \times 10^6}{D^2} \qquad PPSF = \frac{166.277}{D^2}$$

Based on CW measured in millimeters or inches, the speed formula becomes:

$$MPS = \frac{CPM \times 1000}{PPSF \times CW} \times 1.2 \qquad FPM = \frac{CPM \times 12}{PPSF \times CW} \times 1.2$$

10.2 BULK HANDLING

Unlike unit handling, where we looked at how many discrete product units passed a point to determine the rate, with bulk handling we look at the volume or weight of product that passes a particular point, either in cubic feet or cubic meters per minute, or more frequently, in pounds or kilograms per minute or tons per hour.

Just as the size and shape of a product are taken into consideration when determining rates for unit handling conveyors, we must also consider the product characteristics when dealing with bulk applications. Is the product of a consistent size or are there lumps? What are the extremes in size, from a particle to the largest lump? Along with strict length, width, height, and weight, in bulk applications we must also look at other physical characteristics. Flowability, the ability of a product's particles to move around each other, is a major player. Also, is the product stringy like long wood shavings that get tangled together or is it sticky so it collects on the various conveying surfaces? CEMA Standard No. 550 describes in detail and classifies material characteristics.

For troughed belt and pipe conveyors, the rate is determined by the rate at which product is added to the conveyor. The limiting factors become the belt width and speed. The belt can hold only so much. At lower speeds physically, there is no room for more product on the belt.

The belt speed is determined based on three criteria: characteristics of the product to be conveyed, the belt tensioning used, and the desired capacity. Let us look at how product characteristics affect the belt speed.

Product characteristics play an important role in choosing belt speed. Some of these might not have been obvious. When dealing with lightweight or powdery product, the speed needs to be kept low enough to prevent or at least minimize dusting. This is particularly true at the loading and discharge points. Fragile product might also require lower belt speed to minimize degradation of the product due to idler impact as well as at the charge and discharge points.

Heavy, sharp product should be carried at more moderate speeds. The higher speed prevents the sharp edges from unnecessarily damaging the belt cover, especially at the charge point. Also, introducing the product so that the chute feeding the conveyor is angled in the direction of travel will minimize belt damage.

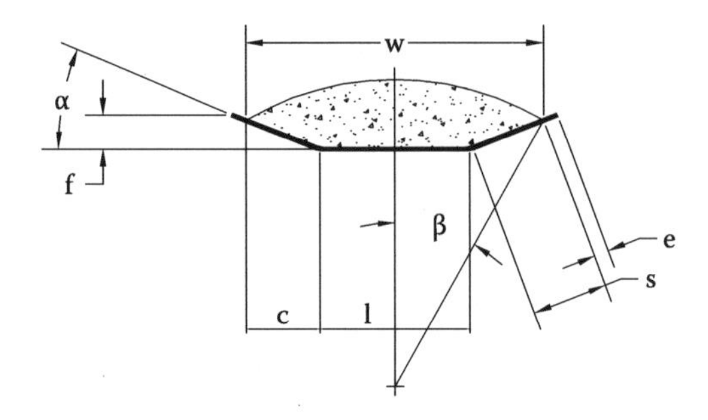

FIGURE 10.4 Troughed belt loading.

When conditions are favorable, speeds can be increased. Increased speed allows for a decreased belt width and decreased belt tension while still achieving the same rate. Speeds can also be increased when dealing with products such as damp sand, soil with no large lumps, and crushed stone. The higher speed does, however, have a trade-off in increased belt and idler wear as well as increased windage-loss from dusting at transition points and product degradation.

Capacity is calculated based on the shape of the product piled on the belt. This requires a fair amount of geometry. Let us look at the calculations. In Figure 10.4 the variables are as follows:

β = angle of surcharge
α = angle of idler
s = slant length
l = length of center idler
e = distance from edge of belt to material
b = belt width
c = horizontal length of slant
w = width of fill area
r = radius of surcharge area
v = belt speed (m/sec)
d = material density (kg/m^3)

Note that all linear dimensions are given in millimeters.

$$e = 0.055b + 23\,\text{mm}\left(0.9''\right)$$

$$c = s \times \cos\alpha$$

$$l = 0.371b + 6\,\text{mm}\left(0.25''\right)$$

$$f = s \times \sin\alpha$$

$$w = l + 2c$$

$$s = 0.2595b - 26\,\text{mm}\left(1.025''\right)$$

$$\text{Area of Rectangle}\left(\text{DEFG}\right) A_R = l \times f$$

$$\text{Area of Both Triangles}\left(\text{ADG}\right) \& \left(\text{BEF}\right) A_T = f \times c$$

$$\text{Area of Surcharge Area}\left(\text{ABED}\right) A_S = r^2\left(\frac{\pi\beta}{180} - \frac{\sin 2\beta}{2}\right)$$

Note that for flat belts, A_R and A_T are zero.

Total area in square meters $A = (A_R + A_T + A_S) \times 10^{-6}$
Volume in kilograms per hour (kg/h) $V = A \times v \times 3600 \times d$

To convert this to metric tons per hour (mtph), simply divide by 1,000.

See Appendix for an abbreviated list of materials and their respective densities.

To simplify this process, CEMA (1994) publishes charts that cover the vast majority of applications with tables for flat belts as well as 20-degree, 35-degree, and 45-degree troughed belts in their book *Belt Conveyors for Bulk Materials*.

For screw conveyors, the volume conveyed is a function of the diameter of the screw as well as the diameter of the shaft. Screw conveyors are not meant to be operated under normal circumstances with product over the shaft and hanger bearings. This can shorten the life of the bearings dramatically. Typically screw conveyors are filled no more than 45 percent. The formula for capacity, measured in cubic feet, uses the RPM of the screw (N), the diameter of the screw (D_S) and the pipe or shaft (D_P), both measured in inches, the pitch of the screw (P) measured in inches, and the percent loading (K).

$$Cap = \frac{N \times 0.7854\left(D_S^2 - D_P^2\right) \times P \times K \times 60}{1728}$$

This formula gives the capacity per hour at a given RPM (N). We must keep in mind that this cannot exceed the maximum recommended speed. To determine the RPM required (N_R), we can change the formula slightly to yield the capacity per RPM and then divide the needed throughput (Cap_R) by that new number.

$$N_R = \frac{Cap_R \times N}{Cap}$$

These formulas are based on the typical application where the pitch of the screw is equal to the diameter of the screw. If the pitch is different, the capacity will be increased or decreased based on a capacity factor (CF). This can be determined using the screw diameter (SD) and the pitch (P), both measured in inches.

$$CF = \frac{SD}{P}$$

If the screw conveyor has a special type of screw such as cut flights, cut and folded, or ribbon flights, the conveyor manufacturer can give you the CF for the specific screw to be used.

Bucket elevators are relatively simple. The capacity in kilograms per hour is based on the bucket volume (V) in liters, the bucket pitch (p) in buckets per meter, the chain or belt speed (v) in meters per second, and the material density (d) in kilograms per cubic meter:

$$Cap = V \times p \times v \times d \times 3600$$

To convert this to metric tons per hour, as before, simply divide by 1,000.

See Appendix for an abbreviated list of materials and their respective densities.

For the rates for the other miscellaneous conveyors that have not been covered, please contact the conveyor manufacturer.

10.3 QUESTIONS

1. What would be the case speed for a system running 30 24″ long cases every minute?
2. If a singulating accumulation conveyor with 24″ accumulations zones is running at 120 FPM, how many 16″ long cartons should be discharged in a minute?
3. What would be the PPSM (product per square meter) of a conveyor 300-mm wide with 50-mm diameter round cans on it?
4. A 15-degree incline has a power feeder with a 5 percent speed differential with the incline its feeding. Does it create a sufficient gap for 30″ long x 30″ tall boxes?
5. What is the recommended minimum speed in FPM for a transportation conveyor carrying 27″ long totes at a rate of 350 totes per hour?
6. Given a live roller conveyor that is 24″ wide with pop-up transfers to sort boxes, what is the required gap if both the conveyor and the transfers are running at 120 FPM and a cycle time of 0.5 seconds?
7. How many metric tons per hour of Portland cement can be moved with a troughed belt conveyor running at 1.5 meters per second? The conveyor belt is 900-mm wide riding on idlers that are 300-mm wide x 20 degrees.

   angle of surcharge = 20°
   angle of idler = 20°
   length of center idler = 300 mm
   distance from edge of belt to material = 72.5 mm
   belt width = 900 mm

belt speed = 1.5 m/sec = belt speed (m/sec)
material density = 1440 kg/m^3 (average for Portland cement)

8. One concrete plant manufacturer builds their screw feeders using 12″ schedule 40 pipe with a 12″ diameter screw on a 2.5″ diameter center shaft. What is the capacity in cubic feet per minute if the screw is running at 240 RPM?
9. If the manufacturer of a shoe sorter stated that the minimum gap required is 6″ and the average carton length is 24″, what would be the throughput rate if the sorter was running at 300 FPM?
10. What happens to the rate in the above questions if the average carton length drops to 18″?

11 Integration and Control Systems

Before selecting conveyors or trying to design a system, it is important to confirm the justification for the system. The four primary sources for system justification are as follows:

1. Cost reduction by increased productivity.
2. Waste reduction through improved handling.
3. Increased capacity through improved space utilization, more output per person, or less downtime.
4. Improved working conditions by reduced manual intervention, reduced injuries, and improved ergonomics.

11.1 SYSTEM DESIGN

Developing a conveyor system is very much like putting together a model railroad. You have a selection of standard components that can be put together in a variety of arrangements to get the train from point A to point B. The shortest distance between two points is a straight line. Straight lines, however, are not always practical; so, the best way is the shortest route that avoids physical obstacles.

The steps for designing a system are fairly straightforward. Start with a layout of the planned installation area and the complete list of products to be handled. Determine the points where the conveyor must go and the areas that it must avoid. Show all the work cells, loading and unloading zones, paint booths, dip tanks, case packers, robots, and other processing equipment with which the conveyor must interface. Be sure to identify areas where the conveyor cannot go, such as through existing process areas, pedestrian areas, the maintenance area reserved for a machine, or through an environment that could have an adverse effect on the conveyor or the product being conveyed.

Select standard equipment whenever possible. It will be less expensive, delivered faster from the manufacturer, and easier to maintain over time.

Now you can draw a preliminary conveyor path. Make sure that the conveyor connects all areas in the proper sequence. Keep in mind that you can go up and over or around obstacles. In some systems, you will have to handle multiple products. For instance, in a polyethylene terephthalate (PET) plastic bottle manufacturing line, you have the following products to handle:

- The plastic pellets that is fed to the injection molder via screw conveyors and/or blowers to produce the preform.
- The plastic preform that is fed into the blow molder typically via a cleated belt conveyor.

DOI: 10.1201/9781003376613-11

- The finished bottle as it exits the blow molder.
- The empty cardboard carton from the exit of the case erector, through the partition inserter, to the case packer. Once filled with the bottles, it is then conveyed to a carton palletizer.
- The palletized load of cartons on a wooden pallet that has to go through a strapper and stretch wrapper before going into the warehouse or onto a truck for delivery to the bottling company.

Because conveyor systems are a significant capital investment, it is important to make sure that the final solution will be appropriate for several years to come. As we mentioned in Chapter 1, "Introduction," it is also important to keep the overall cost of the system in mind. This includes items such as the cost of spare parts and maintenance over the life of the system.

Now that we have taken a broad view of system design, let us take a look at various types of systems in more detail.

11.1.1 Distribution Centers

Distribution centers are generally broken up into areas: bulk storage, pick modules, kitting, merge, induction, and sortation, and finally truck or trailer loading (see Figure 11.1).

In bulk storage, the product is stored in palletized loads stacked in a pallet rack. Typically, there are no conveyors in this area. With the advent of robot shuttles much higher density storage is available.

There are typically two types of pick modules: full-case and split-case. Full-case pick modules usually use long, heavy-duty, flat belt conveyors for the product to be picked. The belt conveyor is used typically because it is less expensive than a live roller conveyor and requires less maintenance.

Split-case pick modules can be arranged in a number of ways. One of the most popular is the central live roller conveyor with gravity roller wings. An empty box or tote starts on the gravity roller, where an operator adds a label to the box to identify the order that is being picked into it. Product is placed in the box, and when an operator is done with his or her portion, the operator pushes it onto the central conveyor, which carries it through the system until the system transfers it into another picking area.

Kitting is an area where products from several locations come together for consolidation prior to going to the final sorter. Product in kitting is sorted by the system so that all of one order goes to a specific station. Once the product is consolidated, it is released to the final shipping sorter.

The merge is where all of the conveyors from the various pick modules and kitting areas all come together.

The induct further combines the product to be sorted and spaces it out on the conveyor so that it is ready to go through the sorter. Once sorted, the product is either directly loaded onto trucks via trailer loader conveyors, palletized in the truck, or palletized on the loading dock and then transported to the truck by forklift. Product that is not successfully sorted is recirculated back into the sortation system at the merge.

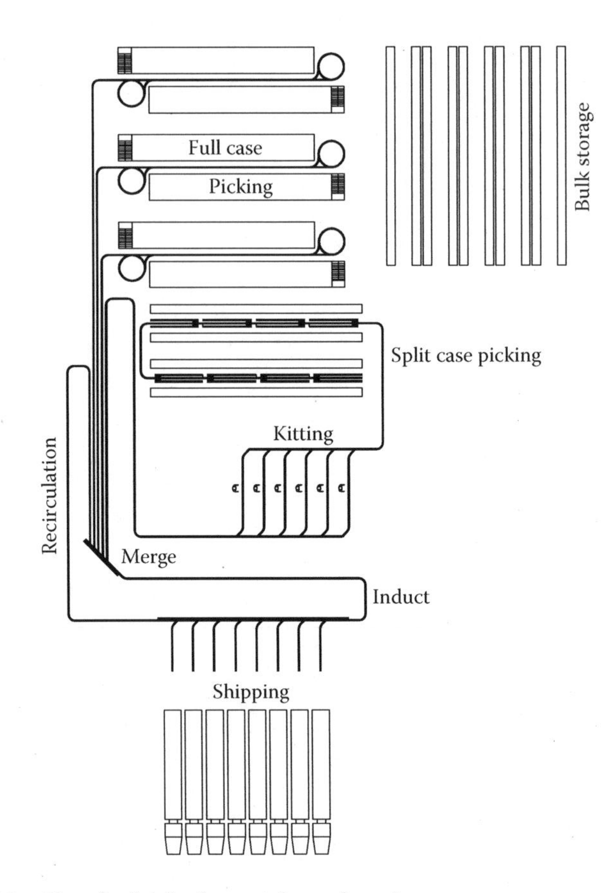

FIGURE 11.1 Generic distribution warehouse layout.

When working on the design for a distribution center, it is best if the conveyor supplier, pallet rack, and forklift supplier can meet with the architect prior to completion of the building design. For one liquor distributor, we were able to work with the architect during the planning stage. Therefore, the columns in the building were spaced in such a way so that they were hidden between the rows of pallet rack and the forklifts had adequate aisle space to maneuver. Also, we were able to make sure the roof structure was sufficient to support any hanging conveyor and that the concrete floor was strong enough to support the point loads. In this case, the architect increased the floor thickness in the areas where we were going to have the legs of a mezzanine.

11.1.2 In-Process Manufacturing

Every in-process manufacturing system is unique. Virtually no two products go through the exact same process, so each system is different. With this type of system, there are typically many more ancillary machines and robots with which to interface. The conveyors should be selected based on their ability to interface with the machines required for the system.

We designed a system for manufacturing rollers for copiers. The system involved a machine to feed the raw tubes onto an indexing chain conveyor. A robot picked the tubes off of the chain conveyor and individually loaded them into a machine that added the end stub shafts. The robot would take the assembled part and place it on a specially designed pallet that would hold six rollers. In all, the system used light-duty chain-driven live roller (CDLR), pop-up chain transfers, a transfer cart, and four more robots. The pallet was subsequently taken to one of several lathes and then to a packaging machine when the rollers were complete. There was even a pallet cleaning station to remove loose metal chips and cutting fluid. This system required a great deal of cooperation among the various equipment manufacturers to ensure successful system integration.

When there is any interface with ancillary equipment, it is important to make sure that the conveyor supplier is in contact with the equipment supplier. Although it is the suppliers' responsibility to provide a working system, it is in your best interests to arrange meetings that include all suppliers and strongly encourage communications. This will greatly reduce the number of interface problems and finger pointing if something does go wrong. Because of this approach, the system just described won productivity awards and the customer's project manager got a promotion for his efforts.

11.2 CONTROL SYSTEMS

Above and beyond the information covered previously, there is a whole profession based solely on the integration of different types of equipment, as well as ancillary equipment such as palletizers, case packers, inspection equipment, barcode scanners, and robots.

Just as people are being asked to do more and varied tasks, the same holds true for conveyors and their control systems. The systems have increased significantly in complexity and sophistication.

The key to any control system is to deliver the product to the right place at the right time in the right condition. This can be as simple as turning on the conveyor and letting it run. In this case, all you need is a power source and a manual motor starter. That is the simplest control system. Typically, to protect the motor from being overloaded and burning out, each motor starter is accompanied by a motor overload switch, a current-sensing device that will shut the motor off if the electrical current drawn by the motor is too high. It works just like a breaker or fuse in your house is designed to protect the wiring from getting overloaded and starting a fire.

If there are multiple conveyors in a line, one feeding the next, we need a way to determine if one conveyor is off so the conveyor feeding it does not keep running. This is done with interlock wiring, a couple of wires that connect each motor overload to the previous or upstream motor starter. This is the simplest integrated control system.

Likewise, if a conveyor is feeding a machine, we need to make sure that the machine is capable of receiving more products. Typically, most industrial machinery comes equipped with specific interlocks. These are usually some wiring terminals in the machine's control cabinet that will tell the feeding conveyor when to run.

This is where we need to pause and discuss control system safety. Each and every control device and machine interlock or interface must be designed to "fail safe." This means that when an emergency stop (e-stop) mechanism is activated, a wire is broken, or a control signal is lost for any reason, the machines and conveyors react in safe mode, a manner that is safest for the equipment and most important, for the personnel involved.

Relays are electrical devices that, when energized, force contacts to shift from their open state to closed and from their closed state to open. Therefore, if some device should be deenergized during a fault condition, it should be wired to the normally open contacts. This way when the power is on, the contacts are closed and the device operates. If, on the other hand, the relay fails, a wire comes loose, or the system is shut off, the contacts will open and the equipment will be safe.

We cover some material now that might seem rather technical. If you do not fully understand it, that is okay. We want to illustrate some of the thinking that needs to go into a control system.

If a device is controlled by a solenoid valve, either hydraulic or pneumatic, it is important to make sure it is fail-safe. The solenoid valve is the control device that determines which lines are under pressure, which lines are exhausted, and which lines are plugged. If the device moves in a vertical path, then a three-position double solenoid with a spring return to center should be used. The solenoid would be energized on one side to raise the device, then energized on the other side to lower the device with the center position if all lines are plugged. This prevents the device from coasting down when deenergized. If it moves in a horizontal path, the solenoid should have an open-center position so that when necessary the device can be moved manually in the case of an e-stop with someone caught in the device. These are just some of the situations that must be considered.

Although you as the customer will not typically be designing the control system, you do need to ask questions about how the system will react to e-stops or various types of failures.

As mentioned earlier, control systems have increased in complexity and sophistication. At the heart of most control systems is a programmable logic controller (PLC). The PLC is a microprocessor-based device that is driven by a program typically written in ladder logic or some version of "C" language. PLCs are designed for use in industrial environments. A PLC can have photo-eyes, limit switches, push buttons, temperature sensors, or other devices connected to inputs and a motor starter, lights, horns, or other devices connected to outputs. PLCs can also interface with personnel through switches, pilot lights, and human–machine interface (HMI) panels.

HMIs are very commonplace in today's automated conveyor system. An HMI is used for much more than simply replacing pushbuttons and pilot lights. Below is a sample screen from an HMI that shows the system layout. The various sections of the system can be color codes to show their current status. The locations of a jam or e-stop that have been actuated can be shown as well. In the bottom left corner, you'll notice a button, Alarm History. This will take you to another screen that lists all of the faults and errors that have occurred (Figure 11.2).

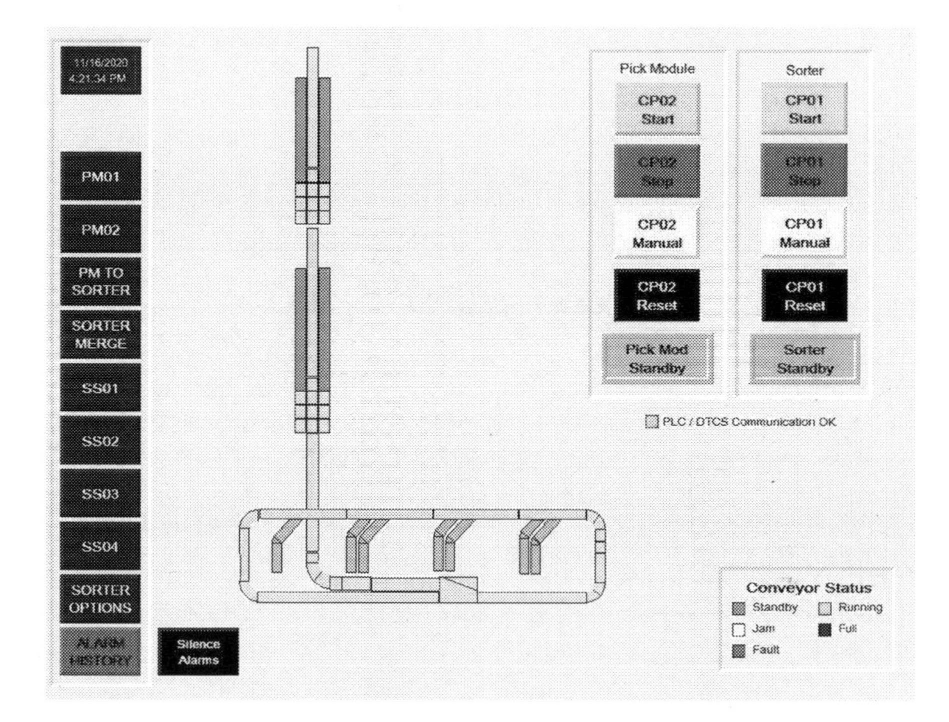

FIGURE 11.2 Sample HMI screen.

PLCs can be programmed to communicate with host mainframe computer systems, warehouse management systems, warehouse control systems, and supervisory control and data acquisition systems (SCADA). They can also be programmed to track system usage and use that information to indicate when periodic maintenance is required. More sophisticated systems are using industrial computers or PCs in lieu of PLCs.

The increased sophistication of control systems is a result of the desire for increased flexibility. Many companies want to be able to use the same conveyor system for multiple products. The changes that can be implemented by the control system include changing conveyor speed, changing product spacing, and in some cases changing the path that the product follows. A conveyor system can be designed to go through a model change between different products.

A system can be designed to handle multiple products simultaneously. The key is to be able to identify each product. This is easily accomplished through the use of radio frequency identification (RFID) tags and tag readers. Each product would have a tag affixed to it, identifying the product. The control system would then know where to route the product through the system.

I would like to offer a warning concerning product identification. A barcode holds a single line of alphanumeric characters, similar to a license plate. With that license plate, you can look up all of the other pertinent information in the host computer system. The same holds true for the two-dimensional bar codes. With RFID tags becoming more prevalent, the inclination is to pack as much information onto the tag as possible. This desire should be resisted when it comes to proprietary information. Innocuous information such as sort lane would be acceptable but do not include things like customer name. Keep the proprietary information in the computer's database, where it is most secure. The data on an RFID tag will stay there until it is erased; therefore, anyone with the proper reader could gain access to the data on that tag. The key is to make intelligent, well-thought-out decisions as to what information to allow out of your facility in that tag.

11.2.1 Controls System Architecture

Depending on the size of the system, one of the decisions you, as the customer, will have to make is whether to use centralized or distributed controls. Centralized controls have been the standard for many years. All control devices such as motor starters, variable frequency drive (VFDs), interface relays, and PLCs are mounted in one central panel. All devices such as motors, photo-eyes, or warning horns must be wired back to the central panel. This was very popular when all of the motors were of high voltage, 480V, and the typical I/O was 110V.

Distributed controls follow a different scheme. The motor starters and VFDs are mounted on the conveyor. Remote input/output (I/O) panels are positioned at various strategic locations throughout the system. The motor starters, photo-eyes, and so on are wired back to the much closer remote I/O panels. Then all that is needed is a simple communication cable that goes back to the main panel where the PLC is located. This concept is less expensive to install from a wiring standpoint.

A hybrid system utilizes a central control panel for motor starters and PLCs and also distributed I/O for all of the lower-voltage photo-eyes, pushbuttons, and warning horns. This is very useful with 24VDC control devices, motorized rollers, and Ethernet communications.

In a conventional, centralized control system, the main panel houses all of the controls. The panel has to be sized to accommodate the PLC, motor starters, overloads, and fuses. This can require expensive heat-dissipating systems as the number of motors increases. A distributed control system eliminates the need for many of these components, thus shrinking the panel size and cost.

With a centralized control, large conduit, and/or cable trays are required to carry all of the wires from the panel to the conveyor drives. As the number of motors increases, so do the size and quantity of the conduit and wireway that are needed. On the other hand, a distributed control system requires minimal wiring due to the close proximity of the motor to its control devices. Reduced wire means cleaner signals in control wires and less interference, a potential source of system malfunction.

Between the three choices, centralized controls are preferred for small or compact systems. Hybrid controls are the correct choice for larger systems that can cover hundreds of meters.

All of these things should be discussed with your supplier to ensure you are getting the best solutions for your application.

We should back up just a moment to pay attention to the 24VDC controls. For many years most photo-eye and pushbuttons were 110V, but 110V can still injure you severely if you get shocked. On the other hand, 24VDC is low enough that the danger is significantly reduced. Another advantage is the wire required for 24VDC lighter gauge and therefore less expensive.

11.2.2 High-Level System Architecture

In many distribution systems as well as production systems, the conveyor control system must communicate the status of the system to a higher-level computer system as we mentioned earlier. Many companies have an ERP (enterprise resource planning) system. While these are great at taking care of orders, invoicing, accounts payable, and accounts receivable, an ERP system frequently falls short when it comes to inventory control. Most offer add-on modules for inventory controls and even warehouse management. To handle the communications between the control PLC and the host computer system, additional software is required. Simple systems will use custom software referred to as middleware. More complicated communications require what is referred to as a WCS (warehouse control system). In an effort to stand out from the next competitor, many companies offer a WES (warehouse execution system) or some other similar acronym. Regardless of what it's called, it must be efficient at handling and conveying information back and forth.

Many of these systems can offer additional reporting features such as:

- Maintaining an error log.
- Track system usage and alert maintenance when it is time to perform periodic maintenance.
- Sending text messages or email to specific people within the organization if certain things occur.
- With the addition of a DVR record video of jams or other issues to aid in problem solving.

The exchange of information can be done in a variety of ways. It can be as simple as the host ERP system writing a text file with all of the information in a pre-agreed-upon format to a specific network location. The WCS can then access that file. The WCS can communicate with the host through the ERP system's API (application programming interface). Another communication option, though less desirable from a data security standpoint, is for the WCS to directly query the ERP system's SQL database.

Another decision is to determine if the servers will be physical or virtual servers. Whether or not they will be physically on the premises (on prem) or cloud-based servers is another decision. The first decision is usually left up to the IT department. The second is frequently more of a corporate decision.

11.2.3 Product Identification

Virtually everyone has been to a grocery store where the cashier scans the barcode on a product to ring it up. There are many different barcode formats. Many have specialized applications that we will not be covering here. Most industrial control systems use Code 128, Code 39, or I 2of 5. Below is a table of the most common barcode formats (Table 11.1).

TABLE 11.1
Barcode Formats

Name	Common Uses		Sample
Code 128	Logistics	Alpha Numeric Characters	Code 128
Code 39		Alpha Numeric Characters	CODE39
Interleave 2 of 5(I 2of5)	Distribution Warehouse	Numeric Characters	01234565
UPC	Retail Product Code	Numeric Characters	1 646265 651114
TUN/ITF-14	Carton Code	Numeric Characters	10123456789019
QR Code	Marketing		
Data Matrix	Aerospace, Automotive, Electronics, USPS		

(*Continued*)

TABLE 11.1 (CONTINUED)
Barcode Formats

Name	Common Uses	Sample
MaxiCode		
PDF417	US Driver's License	

FIGURE 11.3 Sample label.

Some labels contain multiple barcodes. In the label shown below, there is the large 1D barcode that is strictly the number shown below it. There is also a 2D DataMatrix barcode that contains all the readable information on the label as well as the name and address of the recipient and much more (Figure 11.3).

Barcode scanners can differentiate between multiple barcodes on a single box. They can be programmed to only pay attention to barcodes of a specific format, a specific length of characters, or even only code that start with specific letter/number combinations.

Scanners are no longer just the red laser lines that you see at the grocery store. There are image scanners that can record not only the barcodes but also the entire product and record its condition. With the addition of AI (artificial intelligence), image scanners can also identify the differences between bags of kitty and boxes of chew toys. It can also be used for quality control to verify that a tax stamp has been added to a liquor bottle. Another use of the AI-powered image scanner is to determine if a box is damaged, such as crushed or discolored due to wetness.

All of these capabilities should be considered when developing a system or when looking to upgrade an existing system. It goes without saying that as features and capabilities are added the cost will go up as well.

11.3 QUESTIONS

1. What is one reason that Code 128 may be chosen over I 2of5 barcodes?
2. How can you distinguish between multiple barcodes on a single box?
3. When an e-stop is actuated should a vertical lift be allowed to finish coming down or should it stop immediately?
4. What is the value of an open-center solenoid valve for devices that move horizontally?
5. How should an Emergency Stop pull cord switch “fail safe”?
6. Which of the following is a justification for a conveyor system?
   a. Increased capacity through improved space utilization, more output per person, or less downtime
   b. Waste reduction through improved handling
   c. Cost reduction by increased productivity
   d. All of the above
7. How can a SCADA system increase the life of a conveyor system?
8. Which of the following is not a method of data exchange between a WCS and an ERP system?
   a. Utilize the ERP’s APIs.
   b. Text file exchange over the network.
   c. Email exchange.
   d. Directly query the ERP system’s SQL database.
9. What are some of the pros and cons of cloud servers versus “on-prem” servers you can think of?
10. What are some of the advantages of 24VDC controls versus 110VAC?

12 Environmental Considerations

The environment in which the conveyor will be installed will greatly influence some of the design aspects of the conveyor and the system. Is the facility a cleanroom, a warehouse, or a foundry? Is the conveyor outdoors? Is the environment clean, mildly dusty, very dirty, and greasy, or is there wet concrete?

In a dirty environment or outdoors, bearings should have additional seals or shields to extend their life. For example, in a system that was subjected to garnet polishing dust in the air, the bearings had labyrinth seals and were shielded.

In one system, we had lineshaft conveyor running into and out of a machine that filled boxes with wet cement. With wet cement being splashed around, this was a very abrasive environment. Because of the wet cement, very low conveyor elevation, and the various types of drive configurations available for the different types of conveyors, a lineshaft conveyor was determined to be the best choice. The customer understood that the spools would have a relatively short life expectancy and made replacing the spools every year part of the standard periodic maintenance routine.

In a sanitary area, such as a food plant, where the conveyors must be cleaned every day, the motors and control devices must be wash-down rated. The conveyor and supplemental components such as supports must be designed in adherence to U.S. Food and Drug Administration (FDA) requirements.

In a paper plant where toilet paper and paper towels are cut to width, packaged, and boxed, because of the highly flammable paper dust in the air, explosion-proof motors and control devices are used. The same holds true for facilities where printer toner is manufactured, because even though the toner itself is not readily flammable, as airborne dust, it is explosively flammable.

The temperature of a facility is also an important aspect of the environment. As temperatures rise or fall, certain conveyor types need to be modified or cannot be used at all. As a rule of thumb, most conveyors operate satisfactorily at these temperatures:

Powered Conveyors:	+2°C (+35°F) to +40°C (+104°F)
Gravity Conveyors:	–40°C (–40°F) to +121°C (+250°F)

The manufacturer's specifications must be checked any time the environment gets close to the extremes on either end of these ranges. As much as possible, the conveyor system should be designed to minimize the amount of conveyor exposed to temperature extremes. Next, we go through some of the changes that are necessary at various temperatures.

DOI: 10.1201/9781003376613-12

12.1 ABOVE 177°C (+350°F)

Only custom-designed conveyors should be used in this environment.

12.2 BETWEEN +40°C (+104°F) AND +177°C (350°F)

Urethane and polyvinyl chloride (PVC) begin to soften and break down when they reach +60°C (+140°F) whether due to the environment, friction, or a combination of both.

Gravity conveyors, CDLRs, and chain conveyors are the only types of conveyors recommended in this temperature range. The lubricants used in the rollers, bearings, gearboxes, and roller chains need to be specifically developed for this temperature range because normal lubricants break down at elevated temperatures.

Typical air tubing becomes too flexible at elevated temperatures and can come loose from barbed fittings. Compression type or push-in fittings must be used.

12.3 BETWEEN +2°C (+35°F) AND +40°C (104°F)

This is the normal operating range for most conveyors. No modification should be required at these temperatures.

12.4 BETWEEN –7°C (+20°F) AND +2°C (+35°F)

Once an environment gets below freezing, although many of the components will operate satisfactorily, there are certain issues that must be considered. First, any air-operated equipment will require that the air being delivered to it be processed through a dryer to reduce moisture. Compressed air always has some moisture suspended in it, but a refrigerated dryer will remove the vast majority of it. This is important because as the air enters the cold environment, the moisture in it will condense and begin clogging air lines and damaging air-operated devices such as solenoid valves and air cylinders.

The second issue is ice or frost buildup on the conveyor or components. When a conveyor runs between a freezer and a warmer area, anything that comes into the freezer will have warm, moist air around it and the cold will condense that moisture on the conveyor and components. A belt-driven live roller (BDLR) belt would have frost buildup on it and would begin to cause tracking and drive problems. A flat belt conveyor might suffer from the belt sticking to the bed. A chain-driven live roller (CDLR) where the individual chains do not go very far is an excellent choice. Gravity wheel or roller conveyors are typically recommended through a transition because they have no moving components that come in and out of the cold environment.

Other considerations are the following:

- Pulley lagging might have to be changed to maintain sufficient grip on the belt.
- Oversize motors by 50–100 percent.
- Use minimum 150-mm (6-in.) diameter pulleys.

- Avoid nylon materials because they become brittle.
- Consider galvanizing or zinc-plating unfinished steel components to avoid rusting.

12.5 BETWEEN –7°C (20°F) AND –18°C (0°F)

Component selection becomes more important in this temperature range. Pay special attention to belts and lubricant in the bearings and gearboxes. Any equipment normally equipped with an automatic oiler will probably have to be changed to a prelubed chain because the lubricator will not operate properly at these temperatures.

Frost becomes less of a problem in this environment because the air is much drier. Other considerations are as follows:

- All sprockets should have hardened teeth to minimize wear.
- Air cylinders and valves should use silicone rubber seals.
- Belts typically have to change from PVC to urethane covers, as PVC covers will begin to crack.

Keep in mind that the air coming out of the heating, ventilation, and air conditioning (HVAC) system will be significantly colder than the target air temperature for the area. As an example, we were installing conveyor in a freezer that was supposed to be no colder than –18°C (0°F), but the air coming out of the HVAC system was –32°C (–25°F).

12.6 BETWEEN –18°C (0°F) AND –40°C (–40°F)

Component selection becomes critical in this temperature range. At temperatures below –23°C (–10°F), parts that are normally painted should not be painted where a belt will be running over them. The friction between the belt and the paint becomes excessive. This also holds true for areas immediately outside of a freezer door. Use galvanized or zinc-plated steel to minimize belt friction and prevent rust.

In this range, belt capabilities must be verified with the manufacturer. Belts typically have to be changed from urethane to natural rubber.

Urethane lineshaft drive bands should be made from a polyether compound rather than the typical polyester compound.

At –23°C (–10°F) and colder, gearboxes and bearings require special low-temperature lubricants.

Roller chain, regardless of the lubricant, loses its strength at low temperatures. Below –29°C (–20°F), the working load of carbon steel roller chain is only approximately one-fourth of its normal strength.

At –23°C (–10°F) and colder, normal wire and cable insulation becomes too brittle and can crack and fall off, exposing live wires. Electrical contractors must make sure that the wire and cables used have appropriate insulation appropriate for the temperature.

Limit switches and other electrical devices with moving parts will require proper lubrication for the temperature.

12.7 BELOW –40°C (–40°F)

Only custom-designed conveyors should be used in this environment.

12.8 SUMMARY

As you can see, there is quite a bit to keep in mind when subjecting a conveyor to environments beyond the norm. As long as you and your conveyor supplier pay attention to the details, you'll have a safe, functional, and long-lasting system.

12.9 QUESTIONS

1. At what cold temperature should urethane belts no longer be used?
2. Along with paper dust or toner, what other manufacturing plants may require explosion-proof electrical devices?
3. What is the minimum temperature for standard cylinder and solenoid valve seals?
4. If galvanized conveyor beds are recommended for very low temperatures such as below –18°C (0°F), why not use them all of the time?
5. Why would center drives be more popular for belt conveyors used for conveying packages when the temperature is below –7°C (20°F)?
6. At room temperature, what would be the flow characteristics of lubricants used at high temperatures? At low temperatures?

Glossary

Angle of Repose: (n) The angle that the surface of a material in a freely formed pile makes to the horizontal.

Angle of Surcharge: (n) The angle that the surface of a material takes when it is at rest on a moving conveyor belt as measured from the horizontal.

Arguto®: (n) A close-grained, dried, and dense hardwood that is 100 percent impregnated with a blend of nontoxic lubricants. POBCO-B® is another brand name. In operation, the lifetime lubricant is drawn to the working surface and eliminates excessive heat produced by friction. When not in motion, the lubricant is reabsorbed into the bearing.

Average Capacity: (n) The rate averaged over a relatively long period of time. For example, the number of products that go past a specific point in the system during an hour divided by 60 would be the average rate in CPM. Averaging can also be done over a full eight-hour shift. The average capacity is almost always less than the design capacity. This is caused by a variety of conditions such as a lack of product or inefficiencies of workers loading the system.

BDLR (Belt-Driven Live Roller): (n) A roller conveyor where the rollers are driven by a belt.

BOR (Belt on Roller): (adj) Describes a type of belt conveyor where the belt rides on top of a series of rollers.

Bottom Cover: (n) The protective cover on the bottom surface of a conveyor belt, typically rubber, urethane, or PVC.

CEMA: Conveyor Equipment Manufacturers Association, 6724 Lone Oak Blvd., Naples, FL 34109, http://cemanet.org.

Design Capacity: (n) This is the number of product per minute (CPM – cartons per minute) that a conveyor or group of conveyors are designed to deliver.

Dusting: (v) The creation of a cloud of dust around a transition point when dealing with light or powdery bulk materials.

Flowability: (adj) A measure of the ease of individual material particles to move past each other.

Friable: (adj) Crumbly, easily broken.

Gap: (n) The dimension between the trailing end of one product to the leading end of the next product.

Gaylord: (n) Large cardboard boxes attached to a wood pallet.

Guardrail: (n) Rails used to keep products on the conveying surface. This term is used most frequently when referring to package conveyor.

Guiderail: (n) Rails used to guide products on a conveyor. This term is used most frequently when referring to TTC conveyors.

Gravity Conveyor: (n) Term used to describe non-powered conveyors that utilize gravity to keep product moving. Gravity conveyors are also frequently referred to as static conveyors because unless they are at an incline, the product remains in a static location without an outside force.

Head-to-Head: (adj) The dimension from the leading edge of one product to the leading edge of the next product. This can also be referred to as the "pitch" of the product.

HULH: (n) Acronym for heavy unit load handling, generally refers to a class of conveyors that handle individual products that weigh over 200 pounds.

Idler: (n) A type of roller that is not powered, often used to support belts.

Induct: (n) Part of a conveyor system that prepares product for induction into a sorter.

Live Load: (n) The total of all the product on a conveyor divided by the overall length of the conveyor. This number is usually in units of pounds per foot (lb./ft.) or kilograms per meter (kg/m).

Maximum Capacity: (n) The maximum rate that can be achieved on a conveyor or system. This is also referred to as surge capacity and is typically only sustained for very brief time periods.

Monofilament: (adj) A type of belt fabric that uses a single extruded strand of material for the weft. This provides excellent lateral stiffness.

Multifilament: (adj) A type of belt fabric that uses multiple strands of material for the weft. This is used where lateral flexibility is required.

Needle-stitched: (adj) Type of non-woven conveyor belt that has a felt-like surface.

Nose-Over: (n) The transition between horizontal and inclined bed sections on a belt conveyor.

PET: (adj) Polyethylene terephthalate, a plastic used to make many beverage bottles.

PLC: (n) Programmable logic controller, or microprocessor-based control device that can be programmed.

Ply: (n) A layer of fabric or other material in a conveyor belt.

PVC: (adj) Polyvinyl chloride, a plastic used for conveyor belts.

Rate Capacity: (n) This is similar to speed but is a measure of product capacity such as case feet per minute (CFPM) representing how many case feet a conveyor can deliver.

RBO (Radiused Break-Over): (n) The transition between horizontal and inclined bed sections on a belt conveyor. Term originated at Boeing Airport Equipment (BAE) when they were developing baggage handling conveyors.

Self-Cleaning Pulley: (n) A pulley that looks like a 10- or 12-pointed star with the points blunted. This allows anything stuck to the bottom of the belt to navigate the pulley without damaging the belt. Frequently as the belt flexes around the pulley, the debris will come loose and fall off. The center hub of the winged pulley is tapered so that any debris will fall out through the ends of the pulley, thus the name self-cleaning.

Sideboards: (n) Formed metal pans that are attached to the side of a conveyor to keep the product on the conveying surface. This term is most frequently used when referring to troughed belt conveyors.

Sideguards: (n) Formed metal pans that are attached to the side of a conveyor to keep the product on the conveying surface. This term is most frequently used when referring to baggage conveyors.

Singulate: (v) The act of spacing product out in a single file.

Skatewheel: (n) Static wheel conveyor consisting of a series of typical steel wheels similar to those found on children's old roller skates.

Skirtboard: (n) A metal section of a chute used to keep the product centered on the conveyor belt in loading areas. This term is most frequently used when referring to troughed belt conveyors.

Snub Pulley or Roller: (n) A pulley or roller used to increase the amount a belt wraps around a pulley to increase the frictional surface area. It can also be used to aid in tracking the belt.

Speed: (n) This is a distance per unit of time; feet per minute (FPM), meters per second (MPS).

S/R: (adj) Slider/roller, type of belt conveyor, similar to a troughed belt.

TTC: (n) An abbreviation for tabletop chain.

Top Cover: (n) The protective cover on the top of a product carrying surface of a conveyor belt, typically rubber, urethane, or PVC.

Tracking: (v) The process of adjusting idler rollers, pulleys, and loading conditions to insure the belt runs straight.

Training Idler: (n) An idler used on troughed belt conveyors that is mounted on a pivoting base that is equipped with guide shoe or rollers that as the belt pushes against one of them, the idler pivots and steers the belt back to center.

Troughed: (adj) Describes a type of belt conveyor where the belt is held in a concave or troughed shaper by the supporting rollers.

UHMW: (n) An abbreviation for ultra-high molecular weight polyethylene. A plastic material that offers high abrasion resistance.

VFD: (n) Variable frequency drive, a device used to vary the speed of a motor to match process requirements.

Warp: (n) The strands of yarn that run lengthwise in a belt.

Wearstrip: (n) A low-friction material typically used to guide conveyor chains. Wearstrips can be UHMW extrusions, a combination of UHMW extrusion over an aluminum extrusion, or can be machined from solid material.

Weft: (n) The strands of yarn that run left to right or perpendicular to the warp in a belt.

Winged Pulley: (n) See *Self-Cleaning Pulley*.

Appendix
Material Density

The material densities provided below are not an exhaustive list but merely a sampling. In many cases, there are a range of values that can be due to a number of factors. Please use empirical data whenever possible.

Material	kg/m³
ABS resin, pellet	721
Alfalfa, ground	256
Almonds, shelled	481–561
Aluminum oxide	961–1,602
Ash, coal, damp	721–801
Ash, coal, dry	561–721
Baking powder	641–721
Baking soda	1,121–1,281
Barley, flour	400–481
Barley, ground	400–481
Barley, kernel	561–641
Barley, malted	497
Bauxite, crushed	1,201–1,362
Beans, castor	577
Beans, coffee	352–641
Beans, lima	721
Beans, navy	769
Beans, soy	721–753
Bicarbonate of soda	657
Bran, oat	400
Bran, wheat	240–320
Brewer's grain	432
Brewer's grits	529
Brick	1,762
Bronze chips	481–801
Buckwheat	545–673
Buckwheat flour	641
Carbon black powder	64.1–400
Carbon black, pellet	320–721
Carbon tetrachloride	50–60
Carbon, granulated, activated	801–961
Carbon, graphite	641
Cashew nuts	513–593
Castor beans	577
Cement powder, Portland	1,362–1,522
Cement, clinker	1,201–1,442
Cereal flake	192
Coal, ground	641

Material	kg/m³
Coal, lump	721–881
Corn, cracked	561–641
Corn, flaked	96.1
Corn, germ	336
Corn, gluten	416–529
Corn, grits	641–721
Corn, ground	481–561
Corn, meal	513–641
Corn, starch	400–561
Corn, sugar, liquid	1,410
Cullet, glass	1,922
Dirt, dry	1,041–1,281
Dolomite, lump	1,410–1,586
Dolomite, powdered	721
Down, goose	16
Feathers, goose	16
Flour, barley	400–3,684
Flour, corn	481–545
Flour, patent	320
Flour, wheat	481–561
Fluorspar	1,442
Glass bead	1,922
Glass cullet crushed	1,922
Hay	80.1–384
Iron chips	2,643
Iron ore	2,403
Iron oxide	2,883
LDPE, polyethylene	561
Lime, hydrated	400–481
Lime, pebble	881–1,041
Lime, quicklime	400–481
Lime, slaked	513
Limestone, crushed	1,362–1,522
Limestone, dust	1,089
Mica	208–481
Milk powder	240–320
Oats	400–561
Oats, bran	400
Oats, ground	400–481
Oats, rolled	384
Paper, shredded	80.1–192
Peanuts, shelled	561–721
Peanuts, unshelled	240–384
Polyethylene pellet	561–593
Polypropylene, pellet	545–577
Polystyrene, expanded beads	24
Polystyrene, pellet	641
Polyvinyl chloride, pellet	769–833
Pumice	641–721
Rice	721–801
Rock crushed	2146
Salt, coarse crushed	721–881
Salt, granulated	1,121–1,281

Material	kg/m³
Sand, damp	1602
Sand, dry	1,281–1,602
Sand, loose	1,442
Sand, rammed	1,682
Sand, silica	1,522
Sand, water filled	1,922
Sand, wet	1,922
Sand, wet, packed	2,082
Sawdust	64.1–192
Silica sand	1,522
Soda ash	481–721
Soybean, whole	753
Steel, chips	2,403
Sucrose – amorphous	1,506
Sucrose – crystal	1,586
Sugar, brown	721
Sugar, raw	881–1,041
Sulfur, crushed	881–1,121
Sunflower seed	577
Talcum powder	64.1–993
Tea leaves	192
Walnut shells, ground	641–721
Wheat, flour	481–561
Wheat, ground	641
Wheat, whole kernel	721–881
Wood flour	240–400
Wood shavings	48.1–160
Woodchips	320–481
Zinc ore	2002
Zinc oxide	160–481

Bibliography

Ambaflex Inc. *SpiralVeyor SV-Series*. 2023. https://www.ambaflex.com/en/products/spiralveyor-sv-series

Bateman Engineering NV. *Project profile: The Zisco Project*. 1998. http://www.batemanengineering.com/TECHNOLOGY/EngineeringProjectProfiles/Zisco.pdf

Best Conveyors. *Gravity conveyors*. 2007. http://www.bestconveyors.com/Gravity_Conveyors.htm

Conveyor Equipment Manufacturers Association, Ed. 1988. *Screw conveyors*. Naples, FL: Conveyor Equipment Manufacturers Association.

Conveyor Equipment Manufacturers Association, Ed. 1994. *Belt conveyors for bulk materials*. Naples, FL: Conveyor Equipment Manufacturers Association.

Conveyor Section. 2003. *Sortation material handling is handling material*. Charlotte, NC: Material Handling Industry of America.

Conveyor Section. 2006. *Sortation systems, equipment and systems design issues*. Charlotte, NC: Material Handling Industry of America.

Dematic. *Cross belt sorter Dematic S-C 100—High-speed transport and sorting*. 2008a. http://www.dematic.com/32272/40547/Portfolio/Products/SortationSystems/CrossBeltSorterS-C100/btc_dematic_2pic_left.asp

Dematic. Electronic tilt-tray sorter Dematic S-T 100—A fast and versatile sortation system. 2008b. http://www.dematic.com/32272/40545/Portfolio/Products/SortationSystems/ElectronicTilt-TraySorterS-T100/btc_dematic_pic_right.asp

Demetrakakes, Pan. 2008. Conveyor controls keep moving along. *Food & Beverage Packaging* June.

FEECO International. *Bucket elevators*. 2008. http://feeco.com/ProductsandServices/ProductLine/MaterialHandling/tabid/105/BucketElevators.aspx

Fenner Dunlop. *Conveyor belt construction*. n.d. http://www.fennerdunlopamericas.com/pdf/ConstructionFDA0105.pdf

Intralox, Ed. 2002. *Intralox engineering manual*. Harahan, LA: Intralox, Inc.

Itoh Denki, Ed. 2017. *2017-Catalog*. Wilkes-Barre, PA: Itoh Denki USA.

J & S Conveyors, Ed. 1990. *Product catalog*. Honeoye, NY: J & S Conveyors.

Jervis, B. Webb Company. *Webb chain conveyors*. 2005. http://www.jervisbwebb.com/%5CBrochures%5CBul-9238_Webb_Chain_Conveyor_Components_Catalog.pdf

Jervis, B. *Wide Wing Dog Magic® power & free*. 1999. www.jervisbwebb.com/Brochures/Bul-9183_Wide_Wing_Dog_Magic.pdf

Jorgensen Conveyors. *Magnetic conveyors*. 2008. http://www.jorgensenconveyors.com/products/magnetic.html

Litton Industrial Automation, Ed. 1989. *Conveyor technical manual*. Hebron, KY: Litton Industrial Automation Systems, Inc.

Lodewijks, G. 1991. *Modern belt conveyor systems*. Delft, The Netherlands: Delft University of Technology.

Mannesmann Dematic Rapistan Corp., Ed. 1999. *Application guide*. Grand Rapids, MI: Mannesmann Dematic Rapistan Corp.

Mulani, Ishwar G. 2002. *Engineering science and application design for belt conveyors*. Panshan, India: Madhu I. Mulani.

Regal Rexnord. *Modsort® Divert & Transfer Module*. 2023. https://www.regalrexnord.com/brands/system-plast/products/modsort

Regal Rexnord, Ed. 1993. *Rex® MatTop® chains*. Beloit, WI: Regal Rexnord Corporation.

Robbins Conveyors. *Stacker and overland conveyors*. 2008. http://www.robbinstbm.com/products/conveyors/stacker.shtml

Siemens Dematic, Ed. 2003. *Application guide—Heavy unit load handling (HULH).* Grand Rapids, MI: Seimens Dematic.

Staff, Modern Materials Handling. 2006. Marshall White on the state of pallets. *Material Handling Product News* March 1.

Tramco, Inc. *Bucket elevator, the work horse*. 2008. http://www.tramcoinc.com/WebMembersOnly/PDFs/BucketElevator/All%20Pages.pdf

Valu Guide—Nolu Plastics. Ed. 2000. *Conveyor components, custom plastics, industrial parts.* Rancho Santa Margarita, CA: Solus Industrial Innovation.

Vince Hagan Co. *Concrete Reclaimers & Slump Check.* 2023. https://www.vincehagan.com/concrete-batching-equipment/reclaimers-slump-check/

Whirl-Air-Flow Corporation. *Bulk material handling solutions*. 2008. http://airprocesssystems.com/pdf/waf/capabilities.pdf

Index

Pages in *italics* refer to figures and pages in **bold** refer to tables.

I

J

K

L

M

N

O

P

Q

R

S

Printed in the United States
by Baker & Taylor Publisher Services